SpringerBriefs in Computer Science

SpringerBriefs present concise summaries of cutting-edge research and practical applications across a wide spectrum of fields. Featuring compact volumes of 50 to 125 pages, the series covers a range of content from professional to academic.

Typical topics might include:

- A timely report of state-of-the art analytical techniques
- A bridge between new research results, as published in journal articles, and a contextual literature review
- A snapshot of a hot or emerging topic
- An in-depth case study or clinical example
- A presentation of core concepts that students must understand in order to make independent contributions

Briefs allow authors to present their ideas and readers to absorb them with minimal time investment. Briefs will be published as part of Springer's eBook collection, with millions of users worldwide. In addition, Briefs will be available for individual print and electronic purchase. Briefs are characterized by fast, global electronic dissemination, standard publishing contracts, easy-to-use manuscript preparation and formatting guidelines, and expedited production schedules. We aim for publication 8–12 weeks after acceptance. Both solicited and unsolicited manuscripts are considered for publication in this series.

**Indexing: This series is indexed in Scopus, Ei-Compendex, and zbMATH **

More information about this series at http://www.springer.com/series/10028

Zhidan Liu · Kaishun Wu

Mobility Data-Driven Urban Traffic Monitoring

 Springer

Zhidan Liu ⓘ
College of Computer Science
and Software Engineering
Shenzhen University
Shenzhen, China

Kaishun Wu ⓘ
College of Computer Science
and Software Engineering
Shenzhen University
Shenzhen, China

ISSN 2191-5768 ISSN 2191-5776 (electronic)
SpringerBriefs in Computer Science
ISBN 978-981-16-2240-3 ISBN 978-981-16-2241-0 (eBook)
https://doi.org/10.1007/978-981-16-2241-0

This Springer imprint is published by the registered company Springer Nature Singapore Pte Ltd.
The registered company address is: 152 Beach Road, #21-01/04 Gateway East, Singapore 189721,
Singapore

Preface

Real-time, accurate, and comprehensive traffic information is of essential importance to urban transportation. Tremendous efforts have been put in efficiently monitoring the traffic conditions in the past decades. Conventional methods rely on deploying intrusive sensing infrastructures, *e.g.*, traffic cameras, or inductive loop detectors, to actively detect traffic conditions. Due to the excessive deployment and maintenance overheads, it becomes prohibitive when adopting such intrusive solutions at city scale, and as a result, the coverage is limited to certain busy road segments or junctions in most cities.

Thanks to the popularity of ubiquitous sensing and Intelligent Transportation Systems (ITS) in recent years, we can gather unprecedented mobility data by exploiting a variety of mobile devices (*e.g.*, smartphones and on-board GPS devices) and Automatic Fare Collection (AFC) devices widely deployed by urban transit systems (*e.g.*, subways, buses, and taxis). Such emerging big mobility data substantially augments the data availability (coverage and fidelity) and also enriches data diversity, so that large-scale and reliable traffic monitoring becomes viable.

Therefore, in this book, we explore the possibility of exploiting mobility data for efficient and effective urban traffic monitoring. In particular, we will introduce the basis of mobility data and the problem statement of urban traffic monitoring in detail, and we further present a framework of mobility data-based urban traffic monitoring, which describes the typical workflow of data-driven traffic monitoring solutions. Then, we present three mobility data-driven urban traffic monitoring approaches, *i.e.*, a compressive sensing-based approach, a dynamic correlation modeling-based approach, and a crowdsensing-based approach. Specifically, the previous two approaches make use of taxi data for traffic monitoring, while the last one exploits the power of bus riders to achieve low-cost traffic monitoring. Finally, we also discuss some possible research directions, *e.g.*, privacy-preserved mobility data sharing, multi-source data enhanced urban traffic monitoring, and parallel computing promoted deep learning-based traffic modeling, to further improve the accuracy and efficiency of urban traffic monitoring at a large scale.

The intended audience of this book shall be the readers, *e.g.*, researchers, students, or even professionals, who are interested in the research areas of ITS, urban computing, big data analytics, and Internet of Things. In addition, this book can serve as a primer for the beginners to gain a big picture of mobility data-driven urban traffic monitoring and could help them understand the basic workflow of exploiting the mobility data to address a real-world problem.

Shenzhen, China   Zhidan Liu
March 2021   Kaishun Wu

Acknowledgements

This book was supported in part by the National Natural Science Foundation of China (under Grant No.61802261 and No.U2001207), the Guangdong Basic and Applied Basic Research Foundation (under Grant No.2020A1515011502 and No.2017A030312008), Shenzhen Science and Technology Foundation (under Grant No.ZDSYS20190902092853047), the Project of DEGP (under Grant No.2019KCXTD005), and the Guangdong "Pearl River Talent Recruitment Program" (under Grant No.2019ZT08X603).

Contents

Acronyms

AFC	Automatic Fare Collection
ANN	Artificial Neural Networks
ARIMA	Auto-Regressive Integrated Moving Average
CS	Compressive Sensing
FFT	Fast Fourier Transform
GAN	Generative Adversarial Network
GPS	Global Positioning System
GPTE	A Graph Processing-based Traffic Estimation System
HDFS	Hadoop Distributed File System
HMM	Hidden Markov Model
ITS	Intelligent Transportation Systems
MLR	Multiple Linear Regression
OD	Origin-Destination
OSM	OpenStreetMap
RDDs	Resilient Distributed Datasets
RIP	Restricted Isometry Property
SVD	Singular Value Decomposition
SVM	Support Vector Machine

Chapter 1
Introduction

Abstract In this chapter, we will introduce the basic concepts of widely available mobility data in the urban city and their boosted applications. Then we will review and discuss the related works on urban traffic monitoring, and especially highlight these works built on mobility data. Finally, we present the organization of this book.

Keywords Mobility data · Transit systems · Trajectory · Urban traffic monitoring

1.1 Overview

Transportation plays an important role in influencing people's daily life. Thus, intelligent and efficient transportation is the common goal for many metropolitan cities. Therefore, it is critical to have an accurate and efficient way of monitoring and managing the traffic conditions and based on that to improve the utilization of transportation infrastructures. Conventional methods in transportation domain rely on intrusive sensing, where people deploy infrastructural devices like magnetic loop detectors [2] or traffic surveillance cameras at roadsides [32] to actively monitor the urban traffic conditions. Installing such intrusive devices incurs substantial deployment and maintenance overhead but can only provide limited coverage at sparse positions. Besides, mobility surveys were traditionally used to understand the travel demands like the origin-destination matrix (i.e., OD matrix) of people for transport operation and planning [15]. Conventional solutions cannot provide us sufficient understanding to urban traffics in real time. Besides, existing solutions often suffer from biased coverage or inaccurate samplings, and may incur huge labor cost.

The public transit systems generate a large volume of mobility data that contain digital footprints of the urban transportation conditions and operations, e.g., location reports from on-board GPS devices of taxis and buses, smartphone reports of bus and train riders, records of automatic fare collection devices (e.g., e-farecard) deployed by all transit types, including trains, buses, taxis, etc. For example, Fig. 1.1 depicts a 15 min snapshot of location reports from about 10,000 taxis in Shenzhen city, China. We see that most major road segments have been covered by the taxis which makes it possible for us to use taxis as roving probes to measure the traffic conditions of

Z. Liu and K. Wu, *Mobility Data-Driven Urban Traffic Monitoring*,
SpringerBriefs in Computer Science,
https://doi.org/10.1007/978-981-16-2241-0_1

Fig. 1.1 Road coverage by location reports of taxis that can serve as the roving traffic probes

the city and make spatial-temporal predictions of urban traffics. As a result, mobility data driven urban traffic monitoring possesses major advantages in its high coverage of the city, low cost in infrastructure deployment and maintenance, and timeliness in data acquisition and analysis. Therefore, data driven urban traffic monitoring has attracted many research interests in recent years. In this book, we will review and summarize the advances in this literature, introduce several novel urban traffic monitoring approaches, and present our outlooks on this research domain.

1.2 Background

In this section, we will introduce the definitions of mobility data and its boosted applications. In particular, we will briefly review the existing mobility data driven urban traffic monitoring approaches.

1.2.1 Mobility Data

Thanks to the popularity of ubiquitous sensing and Intelligent Transportation Systems (ITS) in recent years, we can gather unprecedented mobility data by exploiting a variety of mobile devices (e.g., smartphones and on-board GPS devices) and automatic fare collection (AFC) devices widely deployed by urban transit systems (e.g., subways, buses, and taxis). Such emerging big mobility data substantially augment the data availability (in both coverage and fidelity) and also enriches data diversity, so that large-scale and reliable urban traffic monitoring and prediction become viable. Formally, devices that can track moving objects (e.g., human beings, vehicles, and so on) will generate a series of timestamped spatial location points, which together forms a mobility trajectory.

Definition (*Location point*) A location point is denoted by $p = \{x, y, t\}$, indicating the instant position at latitude x and longitude y of the moving object at time t. □

Definition (*Mobility trajectory*) A mobility trajectory consists of a sequence of time-ordered location points, denoted by $T = \{p_1, p_2, \ldots, p_m\}$. □

Additional information could also be associated with the location points, which is dependent on specific applications. For example, location points reported from the GPS-equipped vehicles could incorporate instantaneous traveling speeds and directions. Due to the rich sensors in modern smartphones, their collected mobility data could include much more rich information, including acceleration forces and rotational forces, ambient air temperature and pressure, illumination, humidity, noise level, signal strength of WiFi access points (or base stations), and so on.

1.2.2 Mobility Data Boosted Applications

Mobility data contain substantial information about our surrounding physical world, and thus have inspired a variety of novel applications for smart cities and timely decision making. We summarize the mobility data boosted applications in the following four main categories.

- **Human mobility modeling**. The wide availability of mobility data provide us an unprecedented opportunity for a better understanding of human mobility patterns. Modeling human mobility mainly focuses on exploring the spatio-temporal characteristics and potential regularities hidden in individual and population trajectories [38]. Various human mobility models are developed to recognize mobility patterns and predict mobility behaviors [13, 52], and thus support a wide range of applications. For example, mobility models are built from people's daily travel trajectories for designing better bike-sharing systems [46] and public transit systems [30]. In addition, mobility data are used for the analysis of Corona-virus spreading, e.g., monitoring and tracking the spreading of COVID-19 [27].
- **Urban traffic**. Mobility data are highly related with urban transportation and traffic, and thus many ITS and urban traffic monitoring applications have been proposed in the past years [60]. For example, Zheng et al. present a series of works to estimate and predict traffic flows [18, 55], traffic volumes of vehicles on the roads [31], and so on. In addition, Chen et al. proposed an online anomalous trajectory detection to discover possible frauds in the taxi service [4]. In particular, various approaches are proposed to estimate or predict road traffic conditions by exploiting the available vehicular trajectories [20, 22, 24, 25, 48, 64]. This book will focus on urban traffic monitoring, and thus we will discuss more related works in the following subsection.

- **Green transport**. Mobility data have also been commonly applied in green transport applications to optimize the network design of public transit systems, personalized navigation, carpooling and ridesharing. For example, Chen et al. present a night bus planning system by analyzing massive taxi trajectories [5]. Ji et al. study the dynamic redeployment of mobile ambulances for protecting citizens' lives from emergent accidents [14]. Yuan et al. propose to enhance people's driving navigation by learning taxi drivers' intelligence from their GPS traces [49]. In addition, the massive mobility trajectory data allow us to design efficient carpooling services [11, 35, 51] and ridesharing systems [8, 19, 29, 37], so as to alleviate traffic congestion and improve the utilization of vehicles.
- **Location based services**. Mobility data are location-dependent, and thus potentially inspire a variety of location based services in our daily life. Mining mobility data generated from social networking services, e.g., Twitter, Facebook, and Weibo, can reveal the periodic movement of individual user and social movement effect [9], which enables applications like point-of-interest recommendations [7, 16, 47] and tourism planning [6, 26]. Moreover, mobility data can be help site selection by mining the frequent trajectories for increasing business profit and public service quality. For example, T-finder [50] can suggest taxi drivers a hotspot or a travel route to meet more potential passengers. Li et al. present an optimal charging station deployment system by leveraging historical electric taxi trajectory data [17]. Based on the mobility data in a city, billboard placement problem has also been explored for maximizing the influence [56, 58].

1.2.3 Urban Traffic Monitoring

In this subsection, we briefly introduce and discuss existing research works, which focus on urban traffic monitoring, in the literature.

Previous works heavily rely on intrusive sensing infrastructures to monitor real-time traffic flows [28], traffic volumes [2], and traffic queues [33]. Due to the excessive deploying and maintenance costs, however, it is prohibitive to widely adopt them at city scale, which largely limits the coverage. Recent studies leverage instant location reports collected from probe vehicles as an efficient alternative for passive traffic monitoring. They utilize these reports to generate traffic map [21, 61], update digital map [3], predict vehicle travel time [62], predict bus services [10], and model human mobility [13, 46, 52]. These works, however, are often limited by insufficient probe sensors due to privacy concerns or insufficient incentives, and thus suffer from limited monitoring coverage.

By exploiting the traffic correlation among different roads, some attempts have been made to recover the missing traffic conditions through regression [45, 59], matrix factorization [43], and tensor decomposition [42]. Subject to the tremendous computations for large scale traffic estimation, these methods explicitly or implicitly model the traffic correlations linearly. In addition, some recent works [31, 53, 54] exploit multi-source data, e.g., taxi trajectory data, loop detector data, truck trajectory data, and etc. to derive the traffic conditions. Specifically, they build one model for each individual data source and integrate these models for multi-source data to obtain the final traffic estimation results. In Chap. 3, we will present our compressive sensing theory based traffic monitoring approach [22] to recover complete traffic conditions from sparse taxi trajectory data.

There exist some works that adopt advanced models for traffic estimation and prediction [36]. They primarily rely on modeling the temporal correlation of traffics for each individual road segment using Support Vector Machine (SVM) [1], Artificial Neural Network (ANN) [28], and Hidden Markov Model (HMM) [44]. These works are mainly designed for specific roads, e.g., expressways, and implicitly assume sufficient amount of data can be collected. Although the emerging deep learning technique can be used for traffic monitoring [28], it introduces a large number of extra parameters to be estimated and thus requires much more training data, which will greatly increase the computation complexity. The enormous computations involved in non-linear models strictly limit their applicability to the large-scale road network. To overcome the challenges of data sparsity and huge computation overhead, in Chap. 4 we introduce a dynamic traffic correlation modeling based traffic monitoring approach [25], which builds on the optimized graph-parallel processing framework.

Aforementioned works heavily rely on the cooperation with particular companies or transit agencies to access the large amount of mobility data, which are prohibitively obtained without permit. Different from these works, some remarkable works turn to exploit the novel crowdsensing paradigm to proactively collect traffic sensing data for urban traffic monitoring [41]. For example, the authors in [40, 57] treat private vehicles as the crowdsensing workers to contribute their mobility data for traffic monitoring. The dramatic proliferation of rich-sensor equipped mobile devices (e.g., smartphones) has enabled mobile crowdsensing [39], which exploits the power of crowds having smart mobile devices to perform location-dependent tasks at scale. The authors in [12, 61] make use of smartphones to design effective traffic monitoring systems, while [23, 34] study the task assignment problem in crowdsensing to improve the traffic sensing quality. In addition, Zhu et al. suggest to schedule the sensing tasks to augment the urban data based sensing [63]. In Chap. 5, we will present a crowdsensing based traffic monitoring approach [21], which encourages participatory efforts from bus riders to collect lightweight mobility data and derive the traffic conditions through careful data processing and analysis.

1.3 Book Structure

The rest of this book is organized as follows. Chapter 2 will formally define the problem of urban traffic monitoring from mobility data, and then introduce our proposed typical workflow for mobility data based urban traffic monitoring. In Chap. 3, we will introduce a compressive sensing based urban traffic monitoring approach, which exploits the sparse mobility trajectory data from roving vehicles to recover the complete and accurate traffic conditions of a city. In Chap. 4, we will present

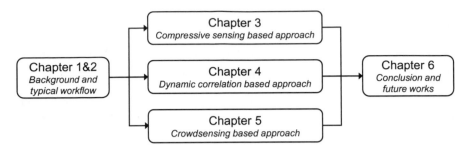

Fig. 1.2 The organization and relationship among the chapters in this book

a more sophisticated traffic monitoring approach, which models the hidden traffic correlation with non-linear models and attacks the involved huge computation challenge with specialized graph-parallel computing design. In addition to mobility data from public transit systems, Chap. 5 will introduce a crowdsensing based urban traffic monitoring approach. In particular, this approach takes advantage of the mobile sensing capability of bus riders to collect lightweight mobility data and recovers traffic conditions through data analysis algorithms. Finally, Chap. 6 concludes the book and will discuss the possible future research directions on mobility data driven urban traffic monitoring.

The relationship among the six chapters is described as Fig. 1.2. More specifically, in this chapter and Chap. 2 provide the background and overview of mobility data driven urban traffic monitoring. Given the proposed framework, Chaps. 3–5 practice the workflow and present three novel urban traffic monitoring approaches. These solutions respectively exploit compressive sensing theory, dynamic correlation modeling, and crowdsensing paradigm to achieve effective and efficient urban traffic monitoring based on mobility trajectory data. Lastly, Chap. 6 concludes this book and points out possible future research directions.

References

1. Asif, M.T., Dauwels, J., Goh, C.Y., Oran, A., Fathi, E., Xu, M., Dhanya, M.M., Mitrovic, N., Jaillet, P.: Spatiotemporal patterns in large-scale traffic speed prediction. IEEE Trans. Intell. Transp. Syst. **15**(2), 794–804 (2014)
2. Aslam, J., Lim, S., Pan, X., Rus, D.: City-scale traffic estimation from a roving sensor network. In: ACM SenSys (2012)
3. Cao, C., Liu, Z., Li, M., Wang, W., Qin, Z.: Walkway discovery from large scale crowdsensing. In: ACM/IEEE IPSN (2018)
4. Chen, C., Zhang, D., Castro, P.S., Li, N., Sun, L., Li, S., Wang, Z.: iBOAT: isolation-based online anomalous trajectory detection. IEEE Trans. Intell. Transp. Syst. **14**(2), 806–818 (2013)
5. Chen, C., Zhang, D., Li, N., Zhou, Z.-H.: B-Planner: planning bidirectional night bus routes using large-scale taxi GPS traces. IEEE Trans. Intell. Transp. Syst. **15**(4), 1451–1465 (2014)
6. Chen, Z., Shen, H.T., Zhou, X.: Discovering popular routes from trajectories. In: IEEE ICDE (2011)

7. Cheng, C., Yang, H., Lyu, M.R., King, I.: Where you like to go next: successive point-of-interest recommendation. In: IJCAI (2013)
8. Cheng, P., Xin, H., Chen, L.: Utility-aware ridesharing on road networks. In: ACM SIGMOD (2017)
9. Cho, E., Myers, S.A., Leskovec, J.: Friendship and mobility: user movement in location-based social networks. In: ACM SIGKDD (2011)
10. Gong, Z., Du, B., Liu, Z., Zeng, W., Perez, P., Wu, K.: SD-seq2seq: a deep learning model for bus bunching prediction based on smart card data. In: IEEE ICCCN (2020)
11. He, W., Hwang, K., Li, D.: Intelligent carpool routing for urban ridesharing by mining GPS trajectories. IEEE Trans. Intell. Transp. Syst. **15**(5), 2286–2296 (2014)
12. Hu, S., Su, L., Liu, H., Wang, H., Abdelzaher, T.F.: Smartroad: smartphone-based crowd sensing for traffic regulator detection and identification. ACM Trans. Sens. Netw. **11**(4), 1–27 (2015)
13. Isaacman, S., Becker, R., Cáceres, R., Martonosi, M., Rowland, J., Varshavsky, A., Willinger, W.: Human mobility modeling at metropolitan scales. In: ACM MobiSys (2012)
14. Ji, S., Zheng, Y., Wang, Z., Li, T.: A deep reinforcement learning-enabled dynamic redeployment system for mobile ambulances. Proc. ACM Interact. Mob. Wearable Ubiquitous Technol. **3**(1), 1–20 (2019)
15. Kuwahara, M., Sullivan, E.C.: Estimating origin-destination matrices from roadside survey data. Transp. Res. Part B: Methodol. **21**(3), 233–248 (1987)
16. Li, H., Ge, Y., Hong, R., Zhu, H.: Point-of-interest recommendations: learning potential check-ins from friends. In: ACM SIGKDD (2016)
17. Li, Y., Luo, J., Chow, C.-Y., Chan, K.-L., Ding, Y., Zhang, F.: Growing the charging station network for electric vehicles with trajectory data analytics. In: IEEE ICDE (2015)
18. Liang, Y., Ouyang, K., Jing, L., Ruan, S., Liu, Y., Zhang, J., Rosenblum, D.S., Zheng, Y.: Urbanfm: inferring fine-grained urban flows. In: ACM SIGKDD (2019)
19. Liu, Z., Gong, Z., Li, J., Wu, K.: Mobility-aware dynamic taxi ridesharing. In: IEEE ICDE (2020)
20. Liu, Z., Huang, M., Ye, Z., Wu, K.: DeepRTP: a deep spatio-temporal residual network for regional traffic prediction. In: IEEE MSN (2019)
21. Liu, Z., Jiang, S., Zhou, P., Li, M.: A participatory urban traffic monitoring system: the power of bus riders. IEEE Trans. Intell. Transp. Syst. **18**(10), 2851–2864 (2017)
22. Liu, Z., Li, Z., Li, M., Xing, W., Lu, D.: Mining road network correlation for traffic estimation via compressive sensing. IEEE Trans. Intell. Transp. Syst. **17**(7), 1880–1893 (2016)
23. Liu, Z., Li, Z., Wu, K.: UniTask: a unified task assignment design for mobile crowdsourcing-based urban sensing. IEEE Internet Things J. **6**(4), 6629–6641 (2019)
24. Liu, Z., Li, Z., Wu, K., Li, M.: Urban traffic prediction from mobility data using deep learning. IEEE Netw. **32**(4), 40–46 (2018)
25. Liu, Z., Zhou, P., Li, Z., Li, M.: Think like a graph: real-time traffic estimation at city-scale. IEEE Trans. Mob. Comput. **18**(10), 2446–2459 (2018)
26. Lu, X., Wang, C., Yang, J.-M., Pang, Y., Zhang, L.: Photo2trip: generating travel routes from geo-tagged photos for trip planning. In: ACM MM (2010)
27. Luo, Y., Li, W., Zhao, T., Yu, X., Zhang, L., Li, G., Tang, N.: Deeptrack: Monitoring and exploring spatio-temporal data: a case of tracking COVID-19. Proc. VLDB Endow. **13**(12), 2841–2844 (2020)
28. Lv, Y., Duan, Y., Kang, W., Li, Z., Wang, F.-Y.: Traffic flow prediction with big data: a deep learning approach. IEEE Trans. Intell. Transp. Syst. **16**(2), 865–873 (2015)
29. Ma, S., Zheng, Y., Wolfson, O.: Real-time city-scale taxi ridesharing. IEEE Trans. Knowl. Data Eng. **27**(7), 1782–1795 (2014)
30. Meegahapola, L., Kandappu, T., Jayarajah, K., Akoglu, L., Xiang, S., Misra, A.: Buscope: fusing individual and aggregated mobility behavior for "live" smart city services. In: ACM MobiSys (2019)
31. Meng, C., Yi, X., Su, L., Gao, J., Zheng, Y.: City-wide traffic volume inference with loop detector data and taxi trajectories. In: ACM SIGSPATIAL (2017)

32. Schoepflin, T.N., Dailey, D.J.: Dynamic camera calibration of roadside traffic management cameras for vehicle speed estimation. IEEE Trans. Intell. Transp. Syst. **4**(2), 90–98 (2003)
33. Sen, R., Maurya, A., Raman, B., Mehta, R., Kalyanaraman, R., Vankadhara, N., Roy, S., Sharma, P.: Kyun queue: a sensor network system to monitor road traffic queues. In: ACM SenSys (2012)
34. Song, S., Liu, Z., Li, Z., Xing, T., Fang, D.: Coverage-oriented task assignment for mobile crowdsensing. IEEE Internet Things J. **7**(8), 7407–7418 (2020)
35. Tong, P., Du, W., Li, M., Huang, J., Wang, W., Qin, Z.: Last-mile school shuttle planning with crowdsensed student trajectories. IEEE Trans. Intell. Transp. Syst. **22**(1), 293–306 (2019)
36. Vlahogianni, E.I., Karlaftis, M.G., Golias, J.C.: Short-term traffic forecasting: where we are and where we're going. Transp. Res. Part C: Emerg. Technol. **43**, 3–19 (2014)
37. Wang, J., Cheng, P., Zheng, L., Feng, C., Chen, L., Lin, X., Wang, Z.: Demand-aware route planning for shared mobility services. Proc. VLDB Endow. **13**(7), 979–991 (2020)
38. Wang, J., Kong, X., Xia, F., Sun, L.: Urban human mobility: data-driven modeling and prediction. ACM SIGKDD Explor. Newsl. **21**(1), 1–19 (2019)
39. Wang, L., Zhang, D., Wang, Y., Chen, C., Han, X., M'hamed, A.: Sparse mobile crowdsensing: challenges and opportunities. IEEE Commun. Mag. **54**(7), 161–167 (2016)
40. Wang, X., Ning, Z., Hu, X., Ngai, E.C.-H., Wang, L., Hu, B., Kwok, R.Y.: A city-wide real-time traffic management system: enabling crowdsensing in social internet of vehicles. IEEE Commun. Mag. **56**(9), 19–25 (2018)
41. Wang, X., Zheng, X., Zhang, Q., Wang, T., Shen, D.: Crowdsourcing in ITS: the state of the work and the networking. IEEE Trans. Intell. Transp. Syst. **17**(6), 1596–1605 (2016)
42. Wang, Y., Zheng, Y., Xue, Y.: Travel time estimation of a path using sparse trajectories. In: ACM SIGKDD (2014)
43. Xin, X., Lu, C., Wang, Y., Huang, H.: Forecasting collector road speeds under high percentage of missing data. In: AAAI (2015)
44. Yang, B., Guo, C., Jensen, C.S.: Travel cost inference from sparse, spatio temporally correlated time series using Markov models. Proc. VLDB Endow. **6**(9), 769–780 (2013)
45. Yang, B., Kaul, M., Jensen, C.S.: Using incomplete information for complete weight annotation of road networks. IEEE Trans. Knowl. Data Eng. **26**(5), 1267–1279 (2014)
46. Yang, Z., Hu, J., Shu, Y., Cheng, P., Chen, J., Moscibroda, T.: Mobility modeling and prediction in bike-sharing systems. In: ACM MobiSys (2016)
47. Ye, M., Yin, P., Lee, W.-C., Lee, D.-L.: Exploiting geographical influence for collaborative point-of-interest recommendation. In: ACM SIGIR (2011)
48. Yoon, J., Noble, B., Liu, M.: Surface street traffic estimation. In: ACM MobiSys (2007)
49. Yuan, J., Zheng, Y., Xie, X., Sun, G.: T-drive: enhancing driving directions with taxi drivers' intelligence. IEEE Trans. Knowl. Data Eng. **25**(1), 220–232 (2011)
50. Yuan, N.J., Zheng, Y., Zhang, L., Xie, X.: T-finder: a recommender system for finding passengers and vacant taxis. IEEE Trans. Knowl. Data Eng. **25**(10), 2390–2403 (2012)
51. Zhang, D., He, T., Zhang, F., Lu, M., Liu, Y., Lee, H., Son, S.H.: Carpooling service for large-scale taxicab networks. ACM Trans. Sens. Netw. **12**(3), 1–35 (2016)
52. Zhang, D., Huang, J., Li, Y., Zhang, F., Xu, C., He, T.: Exploring human mobility with multi-source data at extremely large metropolitan scales. In: Proceedings of the 20th Annual International Conference on Mobile Computing and Networking, pp. 201–212 (2014)
53. Zhang, D., Zhang, F., He, T.: MultiCalib: national-scale traffic model calibration in real time with multi-source incomplete data. In: ACM SIGSPATIAL (2016)
54. Zhang, D., Zhao, J., Zhang, F., He, T.: UrbanCPS: a cyber-physical system based on multi-source big infrastructure data for heterogeneous model integration. In: ACM/IEEE ICCPS (2015)
55. Zhang, J., Zheng, Y., Sun, J., Qi, D.: Flow prediction in spatio-temporal networks based on multitask deep learning. IEEE Trans. Knowl. Data Eng. **32**(3), 468–478 (2019)
56. Zhang, P., Bao, Z., Li, Y., Li, G., Zhang, Y., Peng, Z.: Trajectory-driven influential billboard placement. In: ACM SIGKDD (2018)

57. Zhang, X., Yang, Z., Liu, Y.: Vehicle-based bi-objective crowdsourcing. IEEE Trans. Intell. Transp. Syst. **19**(10), 3420–3428 (2018)
58. Zhang, Y., Li, Y., Bao, Z., Mo, S., Zhang, P.: Optimizing impression counts for outdoor advertising. In: ACM SIGKDD (2019)
59. Zheng, J., Ni, L.M.: Time-dependent trajectory regression on road networks via multi-task learning. In: AAAI (2013)
60. Zheng, Y.: Trajectory data mining: an overview. ACM Trans. Intell. Syst. Technol. **6**(3), 1–41 (2015)
61. Zhou, P., Jiang, S., Li, M.: Urban traffic monitoring with the help of bus riders. In: IEEE ICDCS (2015)
62. Zhou, P., Zheng, Y., Li, M.: How long to wait?: predicting bus arrival time with mobile phone based participatory sensing. In: ACM MobiSys (2012)
63. Zhu, Q., Uddin, M.Y.S., Venkatasubramanian, N., Hsu, C.-H.: Spatiotemporal scheduling for crowd augmented urban sensing. In: IEEE INFOCOM (2018)
64. Zhu, Y., Li, Z., Zhu, H., Li, M., Zhang, Q.: A compressive sensing approach to urban traffic estimation with probe vehicles. IEEE Trans. Mob. Comput. **12**(11), 2289–2302 (2012)

Chapter 2
Urban Traffic Monitoring from Mobility Data

Abstract In this chapter, we will present the problem description of mobility data based urban traffic monitoring, and then elaborate our proposed typical workflow to address this problem. In particular, we describe the common types of mobility data widely available in the urban city, and the ways to collect and process them. Later, we present the typical traffic modeling of mobility data for traffic monitoring.

Keywords Traffic data · Traffic modeling · Traffic speed monitoring · Traffic flow monitoring · Traffic accident monitoring

2.1 Problem Description

The road network[1] of an urban city is composed of a number of roads of different types, e.g., expressways, major roads, minor roads, and etc. Each road is further divided into smaller road segments for better granularity in traffic monitoring. Therefore, a road network can be modelled as a directed graph $\mathcal{G}(\mathcal{V}, \mathcal{R})$, where each vertex $v \in \mathcal{V}$ presents a geo-location (e.g., road intersection), and each edge $r \in \mathcal{R}$ is a road segment. The traffic condition of a road segment r can be measured by the average traffic speed within a time slot. The unprecedented mobility data available in nowadays urban cities contain rich traffic samplings about urban traffics, and provide us an opportunity to implicitly derive the traffic speeds of road segments. A typical traffic sampling includes a timestamp, location, travel speed, and other auxiliary information. For each road segment r, its traffic condition c can be approximated as the average travel speed of all traffic samplings collected at road segment r. The approximation is considered credible if the road segment is sampled by a sufficient number of probe "sensors" [10]. The objective of urban traffic monitoring is thus to derive the *timely*, *accurate*, and *complete* traffic conditions for all road segments in $\mathcal{G}$ based on the available mobility data.

---

[1]Parts of this chapter is reprinted from [6], with permission from IEEE.

© The Author(s), under exclusive license to Springer Nature Singapore Pte Ltd. 2021   11
Z. Liu and K. Wu, *Mobility Data-Driven Urban Traffic Monitoring*,
SpringerBriefs in Computer Science,
https://doi.org/10.1007/978-981-16-2241-0_2

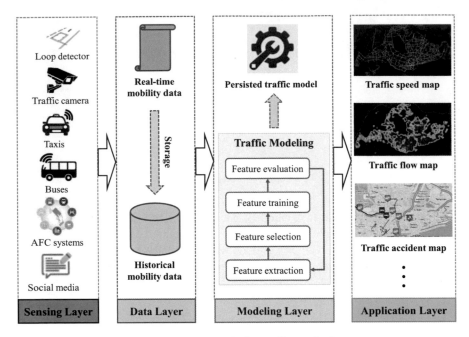

Fig. 2.1   The framework of mobility data based urban traffic monitoring

2.2   Typical Workflow of Urban Traffic Monitoring

Figure 2.1 demonstrates the high-level framework of mobility data based urban traffic monitoring, which includes the sensing layer, data layer, modeling layer, and application layer. Specifically, the sensing layer senses urban traffic and collect mobility data via different devices and infrastructures. The derived mobility data are stored as historical mobility data for model training, while real-time mobility data can be used to estimate the overall traffic conditions. In addition, the modeling layer exploits mobility data to build various traffic models, so as to support urban traffic applications, e.g., generating traffic speed maps, traffic flow maps, or traffic accident maps. In the following subsections, we describe each component in detail.

2.2.1   Sensing Layer

In the urban city, the available mobility data involved in traffic monitoring could be classified into following categories:

Traffic data from infrastructures. Many infrastructure devices, e.g., loop detectors and traffic cameras, have been deployed in the city to continuously collect traffic data. The loop detectors are buried under traffic lanes of some important roads, and can detect the vehicles passing by. Such measurements are used to calculate travel speed of each individual vehicle and also count the total number of vehicles passing by (i.e., traffic flow) within a period. Similarly, traffic cameras are placed above road intersections and used to capture images of vehicles passing by. Based on computer vision techniques, travel speeds of vehicles and traffic flows can also be derived.

Trajectory data from vehicles. In the urban cities, a large number of public vehicles (e.g., taxis and buses) have been equipped with GPS devices, and thus can periodically report their status, including current location, travel speed, travel direction, etc. Those reports indicate the trajectories of vehicles that contain traffic condition measurements of the roads.

AFC records from transit systems. Modern public transportation networks heavily rely on the AFC devices to automatically collect transit fees from both bus and subway passengers, who need to tap their smartcards to AFC readers when they get on and get off the buses or subways. Thus AFC systems record the boarding/alighting (bus or subway) stations/time of passengers, and all of such records can be used to construct a trip OD matrix that reveals mobility flows.

Other data sources. There are some other data sources useful for traffic monitoring. For example, accident reports, which contain location, severity, and event of each accident, are helpful information to assess potential accident risk of each location within a city. Social networking services can treat humans as sensors to probe dynamics of a city and thus the social media data can help to infer traffic anomalies (e.g., accidents) as well. Furthermore, cellphone data indicate users' movements within a city at cell-tower levels, and provide hints for inferences of traffic conditions. In addition, sensing data from crowdsourcing systems also serve as an important data source for traffic monitoring or prediction. All of such data measure urban traffics from a complimentary perspective.

2.2.2 Data Layer

This layer mainly aims to pre-process and store the mobility data collected from the urban city. In general, various pre-processing techniques, e.g., noise filtering and map matching, can be applied to clean and transform the raw data to meaningful information, according to the requirements of specific applications [14]. The processed mobility data then can be stored by some databases for future use, e.g., building traffic models.

2.2.3 Modeling Layer

Urban traffics are complicated and thus some advanced traffic models are preferred, e.g., statistical or machine learning models, to capture the hidden traffic characteristics from mobility data and then facilitate the traffic monitoring or prediction based on the input of real-time mobility data.

As shown in Fig. 2.1, advanced traffic modeling is an iterative process that consists of several phases. To construct a traffic model, we firstly need to extract some desired values (i.e., *features*) from the raw mobility data. Such a set of features are correlated with the target traffic conditions. Taking the traffic condition c_i of a road segment r_i as an example, c_i is not only influenced by traffic conditions of r_i's neighboring road segments on the spatial dimension, but also impacted by time of the day (e.g., peak hour and non-peak hour) and day of the week (e.g., workday and weekend) on the temporal dimension. Those spatial-temporal factors together determine the evolution of c_i and play an important role on accurately predicting its current and even future status. After the feature extraction phase, a small set of the most relevant features are further selected based on some criteria, e.g., information entropy, to simplify the modeling and enhance the generalization capability of a model. After constructing the traffic model only using the most informative and non-redundant features, we can tune the parameters through massive training data and evaluate the derived model with testing data. The whole process of traffic modeling can be repeated until target model performances (e.g., accuracy) are achieved. The persisted traffic model is the one that encodes the traffic characteristics and can be used for traffic estimation or prediction given the real-time input mobility data.

Existing works mainly rely on statistical or machine learning models, e.g., Auto-Regressive Integrated Moving Average (ARIMA) [7], ANN [4], SVM [1], HMM [11], and so on, to capture the complex traffics [9]. When building such traffic models, feature extraction and selection are significantly important as they will determine the final performances of a traffic model. These procedures, however, are heavily dependent on man-crafted feature engineering, which calls for rich experiences and expertise knowledge. Fortunately, the emerging deep learning models [5] could help us automatically determine the best features for constructing an effective traffic model, while these models usually require tremendous computation costs.

2.2.4 Application Layer

Based on the available types of mobility data and the built traffic models, different urban traffic monitoring applications can be developed. Figure 2.1 illustrates the following typical applications for smart cities.

- **Traffic speed monitoring**. Traffic speed is a widely adopted indicator to measure the traffic condition of one road segment, which is generally calculated as the average travel speed of all traffic samplings on a given road segment. Existing works derive such traffic speed measurements either indirectly from data collected by loop detectors [2] and traffic cameras [8] or directly from GPS-equipped vehicles [14]. For a given road segment r_i, its traffic speed can be easily transformed to the travel time [12], since r_i's length is fixed. In addition, some specific applications, e.g., Google Maps, may translate traffic speeds to certain congestion levels (e.g., slow, normal, and fast) according to some mapping rules.

- **Traffic flow monitoring**. In general, traffic flow is defined as the total number of target objects (e.g., vehicles or human beings) that pass through an area during a period. The area can be a road segment or a region in the city. Different from traditional works that hold many assumptions on human mobility, more recent approaches model and estimate traffic flows based on the realistic human mobility data collected from infrastructures and AFC systems. Traffic flows reveal the movements of crowds and potentially determine the traffic distributions [13].
- **Traffic accident monitoring**. Traffic accidents have serious impacts on urban traffic. Therefore, it is necessary to assess traffic accident risks for each specific road or region, which can be measured as likelihoods, meaning how likely is it that traffic accidents might occur on a road/region. Recent practices mainly associate accident risks with current traffic conditions and human mobility, and thus they develop models to mine relations between mobility data and historical accident reports for traffic accident risk estimation [3].

2.3 Summary

In this chapter, we firstly present the problem description of urban traffic monitoring from mobility data, and then propose a generic framework for mobility data based urban traffic monitoring. In particular, the framework includes four layers, i.e., sensing layer, data layer, modeling layer, and application layer, and we elaborate the functions of each layer in detail.

References

1. Asif, M.T., Dauwels, J., Goh, C.Y., Oran, A., Fathi, E., Xu, M., Dhanya, M.M., Mitrovic, N., Jaillet, P.: Spatiotemporal patterns in large-scale traffic speed prediction. IEEE Trans. Intell. Transp. Syst. **15**(2), 794–804 (2014)
2. Aslam, J., Lim, S., Pan, X., Rus, D.: City-scale traffic estimation from a roving sensor network. In: ACM SenSys (2012)
3. Chen, Q., Song, X., Yamada, H., Shibasaki, R.: Learning deep representation from big and heterogeneous data for traffic accident inference. In: AAAI (2016)
4. Jain, A.K., Mao, J., Mohiuddin, K.M.: Artificial neural networks: a tutorial. Computer **29**(3), 31–44 (1996)
5. LeCun, Y., Bengio, Y., Hinton, G.: Deep learning. Nature **521**(7553), 436–444 (2015)
6. Liu, Z., Li, Z., Wu, K., Li, M.: Urban traffic prediction from mobility data using deep learning. IEEE Netw. **32**(4), 40–46 (2018)
7. Pan, B., Demiryurek, U., Shahabi, C.: Utilizing real-world transportation data for accurate traffic prediction. In: IEEE ICDM (2012)
8. Schoepflin, T.N., Dailey, D.J.: Dynamic camera calibration of roadside traffic management cameras for vehicle speed estimation. IEEE Trans. Intell. Transp. Syst. **4**(2), 90–98 (2003)

9. Vlahogianni, E.I., Karlaftis, M.G., Golias, J.C.: Short-term traffic forecasting: where we are and where we're going. Transp. Res. Part C: Emerg. Technol. **43**, 3–19 (2014)
10. Xin, X., Lu, C., Wang, Y., Huang, H.: Forecasting collector road speeds under high percentage of missing data. In: AAAI (2015)
11. Yang, B., Guo, C., Jensen, C.S.: Travel cost inference from sparse, spatio-temporally correlated time series using Markov models. In: VLDB (2013)
12. Yang, B., Kaul, M., Jensen, C.S.: Using incomplete information for complete weight annotation of road networks. IEEE Trans. Knowl. Data Eng. **26**(5), 1267–1279 (2014)
13. Zhang, J., Zheng, Y., Sun, J., Qi, D.: Flow prediction in spatio-temporal networks based on multitask deep learning. IEEE Trans. Knowl. Data Eng. **32**(3), 468–478 (2019)
14. Zheng, Y.: Trajectory data mining: an overview. ACM Trans. Intell. Syst. Technol. **6**(3), 1–41 (2015)

Chapter 3
A Compressive Sensing Based Traffic Monitoring Approach

Abstract Through analysis on real-world mobility data, we observe non-trivial traffic correlations among the traffic conditions of different road segments and derive a mathematical model to capture such relations. After mathematical manipulation, the models can be used to construct representation bases to sparsely represent the traffic conditions of all road segments. With the trait of sparse representation, we propose a traffic monitoring approach that applies the compressive sensing technique to achieve city-scale traffic estimations with only a small number of probe vehicles, largely reducing the system operating cost. Trace-driven experiments with real-world traffic data show that the proposed approach derives accurate traffic conditions with the average accuracy as 80%, based on only 50 probe vehicles' intervention.

Keywords Traffic correlation modeling · Multiple linear regression · Data sparsity · Traffic estimation · Compressive sensing

3.1 Introduction

Understanding traffic[1] conditions is crucial in urban cities, which used to incur heavy road infrastructure constructions, e.g., inductive loop detectors and traffic cameras. Due to the high deploying costs, however, it is prohibitive to densely adopt them in the city scale, which largely limits the coverage. Recent studies leverage roving vehicles on roads as probes to estimate the traffic conditions [9–11]. Vehicles, equipped with GPS, can periodically report their current locations, driving speeds or directions via certain data delivery scheme. A typical traffic sampling contains a time stamp, current GPS position, instant speed and etc. With such on-site traffic information, we can instantly estimate traffic speeds of the roads covered by probe vehicles. Using roving vehicles largely extends the coverage of traffic monitoring. Existing approaches employ extensive probe vehicles to cover the interested roads, while they are often limited by the number of participating vehicles in acquiring the complete

---

[1] Parts of this chapter is reprinted from [6], with permission from IEEE.

© The Author(s), under exclusive license to Springer Nature Singapore Pte Ltd. 2021   17
Z. Liu and K. Wu, *Mobility Data-Driven Urban Traffic Monitoring*,
SpringerBriefs in Computer Science,
https://doi.org/10.1007/978-981-16-2241-0_3

traffic map due to privacy concerns or energy expenditure. A particular road may not have probe vehicles at all times.

Extensive studies report that traffic conditions among road segments are highly correlated [9–11], either in a direct or indirect way. Specifically, the traffic condition of one road is not solely related to any particular individual road but several ones. We refer to such relationship as *traffic correlation* and try to exploit it to achieve complete traffic monitoring at network scale.

3.1.1 Problem Formulation

Given a set of probe vehicles $\mathcal{H} = \{h_1, h_2, \ldots, h_m\}$ and their traffic reports within a time slot, we aim to recover the complete traffic conditions of all road segments. According to the time stamps in the first and last reports from each vehicle within the time slot, we can get their travel time $T = [t_1, t_2, \ldots, t_m]^T$, where t_j indicates the travel time of vehicle h_j. In addition, based on the GPS positions reported by each vehicle, we can recover their travel trajectories on the road network. For each vehicle h_j, we construct a vector $L_j = [l_j^1, l_j^2, \ldots, l_j^n]$, where n is the number of all road segments. Each l_j^i in the vector indicates the distance h_j travels on road segment r_i. When h_j completely travels through road segment r_i, l_j^i equals to the length of this road segment. If h_j only travels a portion of road segment r_i, l_j^i is prorated based on the map matching technique [7]. Based on each L_j, we construct a $m \times n$ matrix $L = [L_1; L_2; \ldots; L_m]$. In general, a statistical average traffic speed vector $S = [s_1, s_2, \ldots, s_n]^T$ of all road segments shall satisfy the following constraint:

$$T = \frac{L}{S} + e, \tag{3.1}$$

where $\frac{L}{S}$ means the element-wise division between L and S. T and $\frac{L}{S}$ may not be exactly equal since S records the average traffic speeds of each road segment. An error vector e is hence introduced in Eq. (3.1).

We can estimate S with the least-square method. Such an estimation approach degrades to mapping all probe vehicles' instant speeds to their traveled road segments if we set the time frame to the exact period interval of traffic samplings. The usual case, however, is that we could not directly calculate S as Eq. (3.1) is usually underdetermined, where the unknowns in S are much more than the measurements in T, i.e., $n \gg m$. Precisely solving such a problem requires at least as many linear independent measurements as the number of road segments.

3.1.2 Compressive Sensing Based Solution

The traffic correlation makes it possible to find a solution to derive traffic conditions for all road segments, due to recent advances in the compressive sensing theory [4]. First of all, we define a traffic condition metric *congestion rate*. Congestion rate c_i of road segment r_i is defined as the reciprocal of its average traffic speed s_i in each time slot, and c_i is formally calculated as follows:

$$c_i = \begin{cases} \frac{1}{s_i} & s_i \neq 0, \\ \infty & s_i = 0. \end{cases}$$

Obviously, smaller congestion rates imply higher traffic speeds and thus better traffic conditions. Hence, we can use average traffic speed s_i and congestion rate c_i to describe the traffic condition of a road segment r_i alternatively. They can be easily calculated from each other.

The compressive sensing theory states that a sparse signal X of size n, i.e., $\|X\|_{l_0} \ll n$, can be reconstructed from m projection measurements, where $m \ll n$ [4]. The measurements are performed using a linear transform Φ on signal X, i.e., $Y = \Phi X$. In our traffic monitoring problem, T is the measurement vector Y, and L is the measurement matrix Φ. The signal $\frac{1}{s} = \mathbf{c} = [c_1, c_2, \ldots, c_n]$, however, is usually not sparse, and we should find an alternative domain Ψ, in which $\mathbf{c}$ can be sparsely represented. It is usually desired to construct such a domain by considering the characteristics in signal X. For our case, traffic correlation in $\mathbf{c}$ is a valuable trait. In domain Ψ, one expects to achieve a k-sparse representation $\mathbf{s}$ for $\mathbf{c}$, i.e., $\mathbf{c} = \Psi\mathbf{s}$, where vector $\mathbf{s}$ contains k non-zero coefficients. As a result, the measurement vector Y can be rewritten as $Y = \Phi\Psi\mathbf{s}$. The compressive sensing theory shows that when $k \ll n$ and measurement matrix Φ and representation basis Ψ satisfy the *Restricted Isometry Property* (RIP) [2], the k-sparse $\mathbf{s}$ can be precisely recovered with high probability by solving the following l_1-minimization problem:

$$\hat{\mathbf{s}} = \arg\min_{\mathbf{s}\in R^N} \|\mathbf{s}\|_{l_1} \quad s.t. \quad Y = \Phi\Psi\mathbf{s},$$

when $m \geq a\mu^2(\Phi, \Psi)k \log n$, where a is a positive constant, and $\mu^2(\Phi, \Psi)$ is the coherence between Φ and Ψ. To further consider the measurements in Y which are usually polluted with error e, i.e., $Y = \Phi\Psi\mathbf{s} + e$ like Eq. (3.1), the equation above can be rewritten as:

$$\hat{\mathbf{s}} = \arg\min_{\mathbf{s}\in \mathbb{R}^n} \|\mathbf{s}\|_{l_1} \quad s.t. \quad \|Y - \Phi\Psi\mathbf{s}\|_{l_2}^2 \leq \epsilon, \tag{3.2}$$

where ϵ is the error tolerance [5]. Equation (3.2) can be efficiently solved by using a standard compressive sensing solver, and thus the global traffic conditions can be recovered as $\mathbf{c} = \Psi\hat{\mathbf{s}}$.

3.2   The Proposed Approach

In this section, we will introduce our proposed compressive sensing based urban traffic monitoring approach, which includes the offline traffic correlation mining and the online traffic estimation.

3.2.1   Traffic Correlation Mining

Traffic correlations among road segments are mined based on historical mobility data. Mining can be performed offline and infrequently to update traffic correlations with relatively long time intervals (e.g., several weeks or months). In the following, we use congestion rates to build the traffic correlation model.

Traffic modeling. Intuitively, the traffic condition of one road segment can be linearly approximated by the traffic conditions of its nearby road segments. We extend such an observation and use the Multiple Linear Regression (MLR) model to capture the hidden correlations among different road segments (even not physically connected) at a global scale. For each road segment r_i, we build the traffic correlation model for its congestion rate c_{r_i} with respect to the traffic conditions of all other road segments in the road network as:

$$c_{r_i} = \beta_{r_i,0} + \sum_{j=1,j\neq i}^{n} \beta_{r_i,r_j} \times c_{r_j} = \mathbf{c_{r_i}}^T \cdot \beta_{\mathbf{r_i}}, \tag{3.3}$$

where $\mathbf{c_{r_i}}$ and $\beta_{\mathbf{r_i}}$ are two $n \times 1$ vectors, and n is the total number of road segments in the road network. The vector $\mathbf{c_{r_i}} = [1, c_{r_1}, \ldots, c_{r_{i-1}}, c_{r_{i+1}}, \ldots, c_{r_n}]^T$ represents the congestion rates of the rest $(n-1)$ road segments except r_i. The element 1 is added to represent the constant item $\beta_{r_i,0}$ in the MLR model. In addition, the vector $\beta_{\mathbf{r_i}} = [\beta_{r_i,0}, \beta_{r_i,r_1}, \ldots, \beta_{r_i,r_{i-1}}, \beta_{r_i,r_{i+1}}, \ldots, \beta_{r_i,n}]^T$ records the corresponding coefficients of each road segment. With sufficient training data, we can estimate the regression coefficient vector $\beta_{\mathbf{r_i}}$ using the least-square method, which minimizes:

$$\sum_{q=1}^{W} \left[c_{r_i}^q - \left(\beta_{r_i,0} + \sum_{j=1,j\neq i}^{n} \beta_{r_i,r_j} \cdot c_{r_j}^q \right) \right]^2,$$

where W is the total number of time frames in training data set and $c_{r_j}^q$ is the congestion rate of road segment r_j in the qth time slot.

A representation matrix can thus be constructed and used to describe the vector $\mathbf{c} = [1, c_{r_1}, c_{r_2}, \ldots, c_{r_n}]^T$, with $(n+1)$ congestion rates, as follows in Eq. (3.4). The element 1 added in $\mathbf{c}$ takes over the constant item in the MLR model.

$$P = \begin{bmatrix} \gamma & 0 & 0 & \cdots & 0 \\ \beta_{r_1,0} & -1 & \beta_{r_1,r_2} & \cdots & \beta_{r_1,r_n} \\ \beta_{r_2,0} & \beta_{r_2,r_1} & -1 & \cdots & \beta_{r_2,r_n} \\ \vdots & \vdots & \vdots & \ddots & \vdots \\ \beta_{r_n,0} & \beta_{r_n,r_1} & \beta_{r_n,r_2} & \cdots & -1 \end{bmatrix}, \tag{3.4}$$

where the ith row corresponds to the MLR model of road segment r_i if we move c_{r_i} to the right-hand side in Eq. (3.3). The coefficients added in the first row is for the constant item in the MLR model and the first element γ is set to be $0 < \gamma < 1$, ensuring that P is invertible. In principle, if our MLR model indeed captures the traffic correlations of all road segments, one would see the projection of c on P, $s = Pc$, to be a vector containing many zero/near-zero entries. In other words, c is transformed to a sparse representation, and the non-zero/significant entries in s capture the major traits of current traffic conditions of the whole road network. If this is the case, P^{-1} serves as the representation basis, i.e., $\Psi = P^{-1}$, which maps s back to c and can be used in compressive sensing technique.

Different traffic scenarios. According to recent studies, urban traffics are not consistent across time and demonstrate different patterns [1, 10]. The underlying traffic correlation may be similar but not identical due to the varying traffics, we thus consider such patterns in our MLR modeling to better capture the traffic correlation, and derive a set of different models for different scenarios. We classify traffic scenarios based on two aspects: different times during a day and different days in a week. We first distinguish traffics of *workdays* from *non-workdays*. We further classify each day (including both workdays and non-workdays) into two periods: period 1 (21:00pm to 07:00am and 13:00pm to 16:00pm: *non-peak hours*) and period 2 (07:00am to 13:00pm and 16:00pm to 21:00pm: *peak hours*). In summary, we consider varied traffic scenarios of four types. Note that the traffic scenario classification is not necessarily fixed to be four, but can be adjusted in different cities.

To train traffic correlation models using actual traffic data set, we first prepare four training data groups according to the time and days. Specifically, we divide the entire time duration covered by the traffic data set into a series of time frames, and map each time frame into one of the four groups, i.e., period 1 (and 2) of workdays (and non-workdays). For each time slot, we compute an average congestion rate based on the associated speeds for each individual road segment. Thus, we have four training data groups with respect to four traffic scenarios at each road segment. Finally, for each data group, we iteratively train the traffic model using MLR in Eq. (3.3) and derive all regression coefficients β_{r_i,r_j} for the matrix in Eq. (3.4). We thus obtain four representation matrices P.

Coefficient pruning. For each road segment r_i, we do not exactly know which set of road segments it actually correlates with before model training. Its congestion rate c_{r_i} in Eq. (3.3), expressed using traffic conditions of all other $(n - 1)$ road segments, can include all possible combinations. In reality, one road segment tends to be more correlated with only a certain number but not all of the road segments, e.g., nearby ones. Correlating all road segments to each particular road segment, the over-fitting issue may occur and huge computation overhead is unnecessarily triggered. Therefore, for any road segment r_i, the traffic correlation would be more reasonably represented by its top κ correlated segments rather than all other $(n-1)$ road segments. We simplify the model in Eq. (3.3) by learning $\kappa + 1$ coefficients merely. To effectively select the top κ correlated road segments for r_i, we consider both the traffic condition similarity and the geographic distance between r_i and any other one road segment r_j. We introduce a selection factor $f_{r_i,r_j} = \frac{d_{r_i,r_j}}{\rho_{r_i,r_j}}$ to differentiate other $(n-1)$ road segments. d_{r_i,r_j} is the geographic distance between the center points of two road segments r_i and r_j, and ρ_{r_i,r_j} is the *Pearson correlation coefficient* of their traffic conditions. A road segment r_j with a smaller f_{r_i,r_j} is possibly more correlated with r_i. After selecting the top κ correlated segments, we prune the number of unknown coefficients from $(n+1)$ to $(\kappa+1)$. For the rest, we simply set their coefficients to 0. In our current implementation, we use the same κ for all road segments in the same traffic scenario, i.e., $(\kappa+1)$ non-zero coefficients in each row of the representation matrix P in Eq. (3.4). A customized selection of κ for each individual road segment (each row in P) will further improve the accuracy. To obtain the appropriate κ values, we may use the accuracy of the derived traffic model to quantify the quality of each κ selection.

Computation complexity. If a road network consists of n road segments and we correlate the traffic condition of each road segment with all other $(n-1)$ road segments, the computation complexity to mine traffic correlations using MLR is $\mathcal{O}(n^3 W)$, where W is the total number of training examples. With the coefficient pruning, however, the computation complexity can be significantly reduced to $\mathcal{O}(\kappa^2 n W)$, where κ is a small constant defined in the "Coefficient pruning" phase and $\kappa \ll n$.

3.2.2   Traffic Estimation via Compressive Sensing

We have constructed the representation bases Ψ for the traffic estimation problem $Y = \Phi \Psi s + e$. In this subsection, we will introduce how to obtain the measurement vector Y and the measurement matrix Φ using a small number of probe vehicles, and then obtain the timely traffic condition estimation of all road segments via a compressive sensing solver.

Measurement vector Y. A server continuously collects traffic samplings from a fleet of probe vehicles in each time slot and estimates the global traffic conditions at the end of the time slot. In other words, the size of time slot is the time granularity of traffic monitoring. A smaller time slot leads to more timely updating of traffic states. In any time slot, supposing the server collects samplings from m probe vehicles, i.e., h_1 to h_m. According to the time stamps contained in the traffic samplings, we get their travel times, and for each probe vehicle h_j, we denote t_{h_j} as its travel time. $Y = [t_{h_1}, t_{h_2}, \ldots, t_{h_m}]^T$ is thus the measurement vector.

Measurement matrix Φ. Based on the GPS positions reported by each probe vehicle, we can calculate their travel trajectories on the road map. To overcome the noise of GPS data, we adopt an hidden Markov model based map matching algorithm [7] to match a sequence of GPS positions sparsely sampled by one probe vehicle to the most likely sequence of road segments. We use the output of map matching as the final vehicle trajectory. Specifically, for probe vehicle h_j, we can construct a vector $L_{h_j} = \{l_{h_j}^{r_i}, i = 1, 2, \ldots, n\}$ according to its trajectory. Each $l_{h_j}^{r_i}$ in L_{h_j} indicates the distance h_j travels on the road segment r_i, which can be calculated as:

$$l_{h_j}^{r_i} = \begin{cases} d_{h_j}^{r_i} & r_i \text{ is passed by } h_j \\ 0 & \text{otherwise,} \end{cases}$$

where $d_{h_j}^{r_i}$ is the actual traveled distance by probe vehicle h_j on road segment r_i. If r_i is fully covered by the trajectory of h_j, then $d_{h_j}^{r_i}$ is the length of r_i; otherwise, the traveled distance is computed as the *great circle distance* via an map matching technique. Based on each L_{h_j}, we can construct an $m \times n$ matrix as $[L_{h_1}; L_{h_2}; \ldots; L_{h_m}]^T$. Considering the constant item $\beta_{r_i,0}$ in MLR model, we add a *zero* value at the first position of each distance vector, and obtain the final $m \times (n + 1)$ matrix as Eq. (3.5) that is the measurement matrix Φ for the compressive sensing formulation. Since each probe vehicle travels freely in the city, the trajectories of any two vehicles are independent [1]. As a result, the measurement matrix Φ constructed in Eq. (3.5) is a random matrix, which satisfies the requirement by compressive sensing on Φ, whose elements should be randomly chosen.

$$\Phi = \begin{bmatrix} 0 & l_{h_1}^{r_1} & l_{h_1}^{r_2} & \cdots & l_{h_1}^{r_n} \\ 0 & l_{h_2}^{r_1} & l_{h_2}^{r_2} & \cdots & l_{h_2}^{r_n} \\ \vdots & \vdots & \vdots & \ddots & \vdots \\ 0 & l_{h_m}^{r_1} & l_{h_m}^{r_2} & \cdots & l_{h_m}^{r_n} \end{bmatrix} \tag{3.5}$$

The compressive sensing theory requires the measurement matrix Φ to be incoherent with the representation basis Ψ for accurate recovery results. In our formulation, the two matrices Φ and Ψ are constructed independently and vary in each time slot. As a result, it is highly possible that they are incoherent.

During the estimation stage, we can partition each vehicle trajectory into multiple shorter pieces to increase the number of independent measurements. After trajectory partitioning, we obtain more traveling time measurements in Y, and accordingly increase the rows in the measurement matrix Φ by splitting each original row in Eq. (3.5) into multiple new rows according to the partitioned trajectories. We find the trajectory partitioning operation does not affect the incoherence between Φ and Ψ, and they are still always incoherent. The natural benefit of trajectory partitioning is that more rows of measurements and constraints are obtained in Y and Φ, thus more traffic details for estimation accuracy. However, the length of partitioned trajectories is generally short. Map matching errors and instant velocity jitters will become obvious and severe. As a result, the benefit outweighs the harm in achieving higher accuracy when we partition the original trajectories into more individual pieces.

Compressive sensing based recovery. We have obtained all the required elements, i.e., Y, Φ, and Ψ, in the compressive sensing formulation. By solving the corresponding l_1-minimization problem of Eq. (3.2) using some standard compressive sensing solvers, e.g., linear programming [3], we can recover $\mathbf{s}$. Then we calculate the congestion rate vector $\mathbf{c} = \Psi\mathbf{s}$, and further obtain the average traffic speeds of all road segments from the reciprocal of congestion rates.

3.2.3 Practical Issues

We discuss two practical issues when we apply our approach for traffic monitoring.

Applying to the larger areas. To control the computation overhead and traffic estimation accuracy, a large city can be partitioned into regions and our approach is then applied to each region. The region can be aligned with the jurisdictional area of each local traffic management bureau. As the connection between two regions is usually based on cascaded roads, the inaccuracy of traffic correlation models for roads along the region boundary is minimal.

Required amount of training data. One concern about the application of our approach to other cities is the size of training data. To learn a MLR model for each road segment in one traffic scenario, the training data size should be $\geq (\kappa + 1)^2$ if we correlate one road segment with κ road segments [8]. Therefore, the minimum training data size would be $W = pn(\kappa + 1)$ for a road network of n road segments with p traffic scenarios. We prefer more training data to weaken the influences of noises and variances of traffic data.

3.3 Experimental Evaluation

In this section, we will perform trace-driven experiments to evaluate the performance of our compressive sensing based urban traffic monitoring approach.

3.3.1 Traffic Data Set

The traffic data set is collected from a fleet of more than 4,400 taxis running in Shanghai city, China. One month traffic data set of all taxis is available for us. Each taxi reports traffic sampling every minute through the cellular network. Equipped with GPS devices in taxis, each reported sampling includes a time stamp, latitude, longitude, instant speed, and other information. The pair of GPS latitude and longitude represents the position where the taxi reports.

Road network. We use a downtown region of Shanghai city, China, as an example to evaluate the performance of our approach. We exploit the OpenStreetMap (OSM)[2] to export all roads in the area, and segment roads according to their intersecting points. In OSM map, two driving directions of the roads are represented separately. Each vehicular trajectory is mapped to the corresponding directional road segments.

As probe vehicles travel along road segments freely in a road network, we thus use the traffic samplings from randomly selected taxis for the traffic estimation to best match the real scenario. By default, we use the traffic samplings from 50 randomly selected taxis for traffic estimation in each time slot. Probe taxis generate adequate traffic data for both model training and performance testing. In total, we have about 5 millions traffic samplings for the testing area.

Traffic data pre-processing. We divide the one month traffic data for each 15 min time slot and classify into four traffic scenarios as described in Sect. 3.2.1. Time slots with size of 15 min are long enough to accumulate sufficient traffic samplings for accurate map matching for each probe taxi, and also provide a fine granularity for timely traffic estimation [10, 11]. We use the data of all taxis in the first three weeks to train MLR models and the data of a small subset of those taxis in the last week to evaluate our approach. In each time slot, we calculate the average traffic speed of each road segment as the average value of all reported speeds by all taxis passing by. In case there is no reported speeds for a road segment in one time slot, we use the average speed of this road segment in the previous 4 time slots. After the traffic data pre-processing, every road segment has a speed in each time slot, and thus the congestion rate. Those average speeds are treated as the *pseudo ground truth*, and used to evaluate the performance of our traffic monitoring approach.

3.3.2 Experiment Results

Traffic estimation accuracy. We first show a snapshot of estimated traffic speeds of all road segments in one randomly selected time slot in Fig. 3.1. This figure shows that the estimated speeds cross different road segments match the trend of the pseudo ground truths very well. To quantify the difference between the pseudo ground truths and the estimated speeds, we plot the CDF of absolute speed differences between them over the four traffic scenarios across the entire 4th-week traffic data in Fig. 3.2a. Due to the inherently higher traffic dynamics in workdays, e.g., the high taxi velocity fluctuation and traffic variance due to traffic jams in workdays, the performance in non-workdays is generally better. The 90-percentile and 60-percentile speed differences are 10.5 and 5.2 km/h for non-workdays and 11.0 and 5.6 km/h for workdays respectively. In general, the estimated speed for each road segment is in good accuracy but always slightly smaller than its pseudo ground truth with overall difference less than 5.0 km/h. It is because in the measurement vector Y, the time difference between the last and the first samplings within a time slot may contain certain time

---

[2]http://www.openstreetmap.org/.

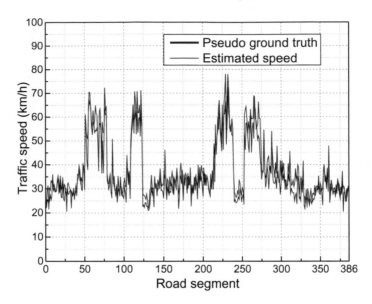

Fig. 3.1 The traffic estimation snapshot of a randomly selected time slot, with comparison of the pseudo ground truths

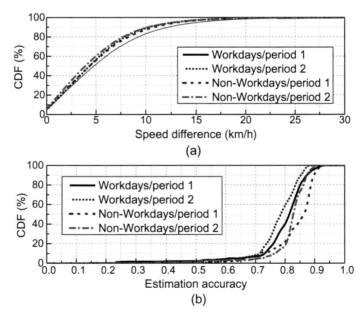

Fig. 3.2 CDF of **a** the absolute speed differences; **b** the estimation accuracy with respect to four different traffic condition indicators

that should not be included, e.g., the waiting time for traffic lights. The measured travel time is thus greater than the actual one and thus a smaller estimated speed.

Instead of directly giving the speed estimations, we translate the estimated speed to a more meaningful traffic indicator for each road. Different from previous methods only providing coarse traffic levels [9], our approach can provide the detailed traffic speed information for each road segment. Similar with Google Map, we classify the traffic conditions of each road segment r_i to four categories according to its traffic speed s_i (in km/h), i.e., *Congested* ($s_i < 20$), *Slow* ($20 \leq s_i < 40$), *Normal* ($40 \leq s_i < 60$) and *Fast* ($s_i \geq 60$). We then compute the estimation accuracy as $\frac{\# \text{ of estimation hits}}{\# \text{ of total time frames}}$, where an estimation hit means both the pseudo ground truth and the estimated speed are classified into the same category in one time slot. We plot the CDF of estimation accuracy over the four traffic scenarios in the 4th-week traffic data in Fig. 3.2b. We find that the four traffic scenarios have similar estimation accuracy distribution and the distribution concentrates within a high accuracy range between 75 and 90%. The accuracy achieves up to 94% and the overall average is more than 80%, which provides quite accurate estimation results.

Compared with a baseline approach. Within a time slot, traffic conditions of the road segments, on which one or more taxis report traffic samplings, can also be measured by the average of instant speeds in traffic samplings, we name such a method as the *baseline method*. Since sparse taxis partially cover a road network in each time slot, baseline method provides an alternative way to obtain traffic conditions of those road segments directly covered by probe taxis only. With sparse probe taxis (i.e., 50 taxis), Fig. 3.3a shows that such an naive approach covers less than 45% road

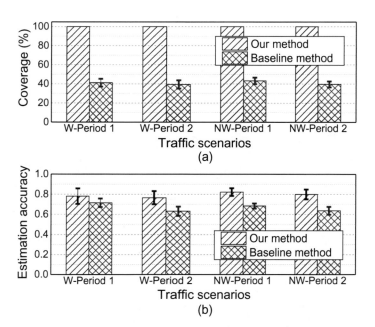

Fig. 3.3 **a** Road coverage by different approaches; **b** Accuracy of roads directly covered by taxis with different approaches

segments of the road network in all the four traffic scenarios, while our approach can always estimate the traffic conditions of the whole road network. In Fig. 3.3b, we compare their estimation accuracy. For a fair comparison, in each time slot, we compare the estimation accuracy only on the road segments with reported traffic samplings (with the same metric as Fig. 3.2b). Due to high variance of each instant speed, the accuracy of the baseline approach is only around 66%. Our approach achieves an accuracy around 80% in all the four scenarios and outperforms the baseline method by 17.13%. We thus adopt the estimated results from our approach for the whole road network. It is worthy to note that our approach uses travel time and distance of taxi trajectory to estimate the average traffic speed of each road segment, and thus avoids the biased speed observations in traffic samplings due to dynamic taxi behaviors.

Compared with an advanced approach. The SVD-TE approach [11] utilizes the spatial temporal correlation in traffic matrix constructed by recent traffic samplings to recover missing elements. Specifically, SVD-TE accumulates traffic samplings in an $n \times t$ traffic matrix and then leverages the Singular Value Decomposition (SVD) technique to recover the missing traffic conditions in the traffic matrix. n is fixed as the total number of road segments in a road network, and t can be varying which determines a window size to accommodate traffic conditions in the SVD recovery. Figure 3.4a shows that the estimation accuracy of SVD-TE is low when the matrix

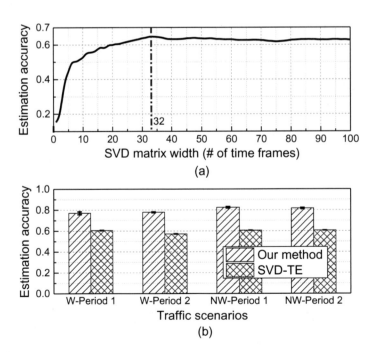

Fig. 3.4 **a** Accuracy of SVD-TE with various matrix width t; **b** Comparison on accuracy of our approach and SVD-TE

width (i.e., t) is small and stabilizes around 0.63 when it is greater than 32 time slots (i.e., 480 min). In Fig. 3.4b, we compare our approach with SVD-TE when t adopts 32. Our approach outperforms SVD-TE on the estimation accuracy by about 20%. In addition, our approach can estimate traffic conditions of the entire road network every 15 min (i.e., one time slot). Such a short delay is necessary in practice due to the timeliness of traffic monitoring applications. The major difference between our approach and SVD-TE is that we have explicitly built the traffic correlation model and thus strengthened efficiency of compressive sensing technique, to which the great advantages of our approach are attributed.

3.4 Summary

This chapter presents a compressive sensing based urban traffic monitoring approach, which is able to achieve city-scale traffic estimations with only sparse traffic probes. The strong correlations among the road network is captured by an explicit model and further exploited to form a space basis that can sparsely represent the road traffic conditions. The trace-driven experiments show that our approach achieves accurate and complete traffic estimations with only sparse probes.

References

1. Aslam, J., Lim, S., Pan, X., Rus, D.: City-scale traffic estimation from a roving sensor network. In: ACM SenSys (2012)
2. Candès, E.J., Romberg, J., Tao, T.: Robust uncertainty principles: exact signal reconstruction from highly incomplete frequency information. IEEE Trans. Inf. Theory **52**(2), 489–509 (2006)
3. Candes, E.J., Tao, T.: Decoding by linear programming. IEEE Trans. Inf. Theory **51**(12), 4203–4215 (2005)
4. Donoho, D.L.: Compressed sensing. IEEE Trans. Inf. Theory **52**(4), 1289–1306 (2006)
5. Donoho, D.L., Elad, M., Temlyakov, V.N.: Stable recovery of sparse overcomplete representations in the presence of noise. IEEE Trans. Inf. Theory **52**(1), 6–18 (2005)
6. Liu, Z., Li, Z., Li, M., Xing, W., Lu, D.: Mining road network correlation for traffic estimation via compressive sensing. IEEE Trans. Intell. Transp. Syst. **17**(7), 1880–1893 (2016)
7. Newson, P., Krumm, J.: Hidden Markov map matching through noise and sparseness. In: ACM SIGSPATIAL GIS (2009)
8. Weisberg, S.: Applied Linear Regression, vol. 528. Wiley, New York (2005)
9. Yang, B., Guo, C., Jensen, C.S.: Travel cost inference from sparse, spatio-temporally correlated time series using Markov models. Proc. VLDB Endow. **6**(9), 769–780 (2013)
10. Yang, B., Kaul, M., Jensen, C.S.: Using incomplete information for complete weight annotation of road networks. IEEE Trans. Knowl. Data Eng. **26**(5), 1267–1279 (2014)
11. Zhu, Y., Li, Z., Zhu, H., Li, M., Zhang, Q.: A compressive sensing approach to urban traffic estimation with probe vehicles. IEEE Trans. Mob. Comput. **12**(11), 2289–2302 (2012)

Chapter 4
A Dynamic Correlation Modeling Based Traffic Monitoring Approach

Abstract This chapter presents a graph processing based traffic monitoring system, GPTE, which is able to achieve high accuracy and high scalability to support city scale traffic monitoring. GPTE benefits from its non-linear traffic correlation modeling and the graph-parallel processing framework built on clustered machines. By representing the road network as a property graph, GPTE decomposes the numerous computations involved in non-linear traffic models to vertices and performs traffic estimation via neural network modeling and iterative information propagation. This chapter presents our experiences in designing and implementing GPTE on top of the Apache Spark. Experiments based on real-world mobility data show that GPTE can achieve the city-scale traffic estimations within 34 s, with the average accuracy as high as 88%.

Keywords Non-linear correlation modeling · Artificial neural network · Graph-parallel processing · Graph partitioning · Information propagation

4.1 Introduction

Probe vehicles based urban traffic monitoring usually[1] suffers from the data sparsity issue, where the available vehicular data cannot fully cover the road network in both spatial and temporal aspects. Existing works primarily exploit traffic correlations among different road segments to recover the complete traffics from incomplete road measurements [6, 17, 21]. The traffic correlations used in those works are explicitly or implicitly modeled linearly mainly to reduce computations for large road networks. The practical traffic, however, is influenced by various factors, e.g., intricate road network, transport regulations, mixed traffic flows, and so on, and thus are much more complex than linearly, which requires advanced modeling for more accurate estimations. There exist a few studies that consider non-linear traffic correlation models [2, 17], all of which, however, suffer from poor scalability due to enormous computation overheads involved in non-linear modeling, which strictly limits their practical applicability to larger scale traffic estimations.

---

[1]Parts of this chapter is reprinted from [7], with permission from IEEE.

© The Author(s), under exclusive license to Springer Nature Singapore Pte Ltd. 2021   31
Z. Liu and K. Wu, *Mobility Data-Driven Urban Traffic Monitoring*,
SpringerBriefs in Computer Science,
https://doi.org/10.1007/978-981-16-2241-0_4

This chapter proposes a non-linear model to characterize the traffic correlations for more accurate traffic estimations. In order to cope with the heavy computations introduced by the model, we develop a graph processing framework that can be efficiently executed in parallel on the computer cluster. In the proposed framework, the road network is represented as a property graph, where vertices are road segments and edges are formed between connected road segments. By distributing the property graph among machines, we decompose all computation tasks at each vertex and perform complete traffic estimations via information propagation among vertices. This solution embraces non-linear traffic correlations for higher estimation accuracy and can be highly parallelized for city-scale traffic monitoring.

The idea being attractive, developing a practical system out of it is challenging for at least two reasons. First, it is non-trivial to model non-linear traffic correlations within a graph. As probe vehicles freely travel across roads, the traffic states of graph vertices are randomly sampled and they form a time-evolving graph. Due to such dynamics, the vertices with known traffic states are changing over time. As a result, it is impossible to maintain a fixed correlation model throughout traffic estimations. Second, a number of system details need to be carefully addressed when implementing the full system. In particular, when we fit our solution into a cluster computing platform, we must not only consider computation cost but also minimize the communication overhead associated with data exchange and computing threads running on different machines. This requires wise treatments to the traffic data given road network structure and characteristics.

To addresses above challenges, we propose *G*raph-parallel *P*rocessing based *T*raffic *E*stimation—GPTE. Specifically, GPTE represents road network data as a property graph and annotates vertex states with real-time traffic samplings. GPTE builds ANN models to capture the correlations and iteratively propagates traffic information from annotated vertices to those vertices of unknown states. To deal with dynamics in the time-evolving graph, for each un-annotated vertex GPTE dynamically selects correlated vertices from its annotated neighbors, and builds an instant ANN model to infer its traffic state. We build GPTE based on the latest cluster computing framework Apache Spark [19], and make use of interfaces provided by the recent graph processing engine GraphX [5]. To reduce communication cost during the cluster execution, we propose a geography-aware graph partitioner that optimizes the data layout on different machines. We improve the efficiency of information propagation among vertices using multi-hop message broadcast scheme and redundant message elimination.

4.2   Non-linearity in Traffic Modeling

There have been many studies that exploit traffic correlation among road segments in order to complete the traffic estimation. The traffic of nearby road segments is mutually influenced, and thus their traffic conditions are highly correlated [6, 17, 21]. Previous works explicitly or implicitly model the traffic correlations linearly mainly

Fig. 4.1 **a** A simple road network. **b** The three-layer ANN model for capturing the traffic correlations between road segment r_3 and r_1, r_2, r_4 and r_5

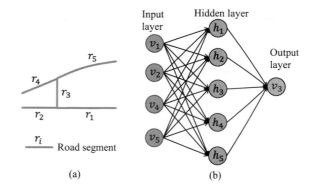

(a) (b)

to reduce the tremendous computation overheads, and recover the complete traffic conditions by exploiting techniques like regression [18, 20], matrix factorization [16], tensor decomposition [13, 15], and compressive sensing [6, 21]. The practical traffic, however, is affected by a number of factors, e.g., intricate road network, traffic regulations, traffic lights, mixed traffic flows, weather condition, etc. and far more complex than linearly.

To better capture inscrutable traffic correlations, some recent works suggest building a non-linear model by ANN [2, 8, 14]. The ANN model is remarkable in representing complex patterns from imprecise data, and enables us to capture detailed traffic correlations for each road segment. Specifically, for road segment r_i and its n correlated road segments, we can build a three-layer feed-forward neural network to model their traffic correlations. The desired output of the model is traffic condition c_i of r_i, while the input layer consists of traffic conditions c_j of each correlated road segment r_j. The hidden layer of the model contains $(n + 1)$ hidden unit h_k.[2] Both hidden and output units can use a standard hyperbolic tangent (i.e., tanh) activation function. Therefore, the ANN model includes $n \times (n + 1)$ input-to-hidden parameters, $(n + 1) \times 1$ hidden-to-output parameters. With sufficient training data of m samples, a standard back-propagation algorithm can be adopted to train the ANN model by minimizing the mean square error as the optimization objective. The computation complexity to train an ANN model is $\mathcal{O}(m * n^2)$.

For example, assuming the correlated road segments of road segment r_3 as depicted in Fig. 4.1a are r_1, r_2, r_4, and r_5, we can build the three-layer neural network as shown in Fig. 4.1b, which includes 4 input units, 5 hidden units, and 1 output unit. We train the model using historical mobility data, and use the derived model to connect r_3's traffic condition v_3 with traffic conditions of r_1, r_2, r_4, and r_5.

Challenges. While ANN based traffic correlation modeling could improve estimation accuracy, it is challenging to introduce the ANN modeling for large scale estimations. First, fine ANN modeling for traffic correlations is non-trivial. A proper model needs to determine the correct correlated road segments, which involves var-

---

[2]We empirically set the number of hidden units in the hidden layer as $(n + 1)$ to balance the computation overhead and prediction accuracy.

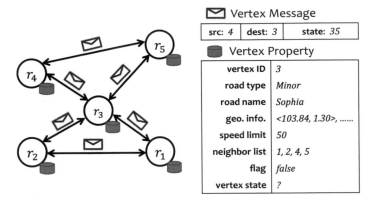

Fig. 4.2 Road network data as a property graph

ious combinations in the road network space and thus results in huge computations. In addition, the random movements of probe vehicles bring uncertain availability of traffic samplings [3], which requires dynamic ANN modeling during the estimation stage. The inherent complexity in building ANN models will inevitably introduce tremendous computation overheads, especially when we scale the traffic estimation to the entire city that involves tens of thousands of road segments and the related ANN modeling. Such computation overheads may overwhelmingly degrade the timeliness of traffic estimations.

4.3 Think Like a Graph

To address the challenges, we present GPTE. In this section, we detail how GPTE represents road network data as a property graph and enables non-linear correlation modeling based traffic estimation on the property graph.

4.3.1 Graph Representation

We model the underlying road network for traffic estimation as a property graph $G(V, E)$, where road segments are represented as vertices and edges are formed between any two physically connected road segments. Figure 4.2 depicts the corresponding property graph for the simple road network presented in Fig. 4.1a. As the traffic conditions of connecting road segments are mutually influenced, the edges are thus bidirectional. In such a property graph, each vertex r_i owns static properties (e.g., road type, road name, geographic information, speed limit c_i^{max}, neighbor list $\mathcal{N}_i^h$ within h hops) and dynamic properties (e.g., flag and vertex state c_i). Specifically, geographic information contains a series of longitude and latitude coordinates to describe the locations of one road segment, with which we can match traffic samplings of probe vehicles to appropriate road segments they traveled on. c_i^{max} is the permissible maximum travel speed on road segment r_i, which can be acquired from the transport agency. $\mathcal{N}_i^h$ stores neighboring vertices of r_i within h hops. c_i is current traffic condition of r_i, and $\texttt{flag}$ indicates whether c_i is known or not. The users can easily incorporate other properties to the vertices if necessary. In graph-parallel processing frameworks, each vertex is able to send (and receive) messages along edges to (from) neighboring vertices. Each message contains source vertex ID, destination vertex ID, and source vertex state c_i. Based on message propagation and local vertex computation with messages, different correlation models and inference algorithms can be implemented on the property graph.

4.3.2 Dynamic Correlation Modeling

Basic ANN modeling. Based on the property graph, a straightforward approach to embedding non-linear traffic correlations into traffic estimation is to pre-learn an ANN model for each vertex[3] and use this model for online vertex state inference. For vertex r_i, we can connect traffic condition c_i with its immediate neighbors in $\mathcal{N}_i^1$ and build an ANN model to capture their traffic correlations for inferring c_i. Ideally these ANN models can be learned offline to lessen the computation burden of online traffic estimations. However, it cannot work in practice due to the dynamics of the property graph. As probe vehicles randomly sample road traffic conditions, the vertices annotated by sufficient probes change from time to time, resulting in time-evolving graph. Such dynamics cause uncertain availability of input vertex states in the ANN models and as a result the fixed ANN model may not work when the vertex states are missing.

    Dynamic ANN modeling. To deal with dynamics in the time-evolving graph, we propose the *dynamic correlation modeling*. Instead of maintaining fixed ANN models, GPTE builds instant correlation models during online traffic estimation stage and performs traffic estimation via iterative message propagation and vertex state inference on the graph. Within each time slot, GPTE associates the traffic samplings from probe vehicles to vertices according to their locations. For a vertex r_i visited by sufficient number, e.g., $\geq \lambda$, of probe vehicles, GPTE annotates its state c_i as the average of all travel speeds and sets $\texttt{flag}$ as *true*. Otherwise, GPTE keeps c_i as un-annotated and sets $\texttt{flag}$ as *false*. GPTE iteratively infers unknown vertex states from annotated vertex states via dynamic ANN modeling, which consists of message propagation, correlated vertex selection, and ANN based vertex state inference. These procedures are repeated for iterations until all vertex states are updated. The final output is the set of all vertex states, which corresponds to the traffic conditions of the entire road network. We detail each procedure in the following.

---

[3]In the following, we use *vertex* and *road segment* interchangeably.

(1) Message propagation. In each iteration, only annotated vertices send messages to neighbors along edges, and all vertices receive messages. Each vertex r_i will record the annotated neighbors in a set C_i based on the received messages. The un-annotated vertices make use of these messages to infer their own states via dynamic correlation modeling. Specifically, for each un-annotated vertex r_i, it will select several most correlated vertices from the annotated neighbors that have sent messages to r_i, and then build an instant ANN model to infer its own state c_i.

(2) Correlated vertex selection. Although we can simply treat all items in C_i as the correlated vertices of r_i to build an ANN model, it may result in poor inference accuracy when considering uncorrelated vertices. To balance the computation overhead and model accuracy, GPTE only selects the most κ correlated vertices from C_i for ANN modeling.

We select correlated vertices for r_i from a data perspective. Specifically, we measure the traffic characteristics of each vertex using its historical traffic data, and determine κ most correlated vertices for r_i according to the correlations of their traffic data. In practice, one vertex may have no direct impact on the target vertex r_i's traffic state c_i, while it may implicitly affect c_i when combined with several other vertices. Identifying such a combination of vertices from the graph space is computationally complex. GPTE adopts a modified feature selection algorithm, i.e., mRMR [11], to efficiently approximate the optimal correlated vertex selection. mRMR is able to maximize the relevance while minimizing the redundancy among selected features.

We treat vertex $r_j \in C_i$ as a *feature* for inferring the state c_i of r_i, and use *mutual information* as the criteria to measure non-linear traffic correlations between vertices. For vertex r_i, we treat its state c_i as a random variable X_i. Thus its entropy is defined as $H(X_i) = -\sum_{x \in X_i} P(x) \log(P(x))$, where x denotes a specific value of X_i, and $P(x)$ denotes the probability of x over all possible values of X_i, which can be calculated from historical traffic data. The mutual information between two vertices r_i and r_j is measured as

$$I(X_i; X_j) = \sum_{x \in X_i} \sum_{y \in X_j} P(x, y) \log \left(\frac{P(x, y)}{P(x)P(y)} \right),$$

where $P(x, y)$ is the joint probability of x and y. $I(X_i; X_j)$ is used to measure the dependency between X_i and X_j. The mRMR algorithm can be further updated by using the mutual information and the selection criterion in Eq. (4.1) to evaluate each un-selected feature [4, 11]:

$$NJ(X_k) = NI(X_k; X_i) - \frac{1}{|\mathcal{F}_i|} \sum_{X_j \in \mathcal{F}_i} NI(X_k; X_j). \tag{4.1}$$

In particular, for a target road segment r_i and the feature set $\mathcal{F}_i$ that has been selected, the first term in Eq. (4.1) measures the mutual information of an un-selected feature $r_k \in C_i$ and target r_i, which is denoted as *dependence*, while the second term measures the average redundancy of r_k and the already selected features in $\mathcal{F}_i$, which is denoted as *redundancy*. These two terms together determine whether r_k should be further included in $\mathcal{F}_i$.

Different from the original mRMR algorithm in [11], we normalize the mutual information $I(X_k; X_i)$ by $H(X_i)$ according to [4] and define $NI(X_k; X_i) = \frac{I(X_k; X_i)}{H(X_i)}$, so that the normalized mutual information $NI(X_k; X_i) \in [0, 1]$, because the entropy of a feature itself could vary greatly. However, $H(X_i)$ cannot be used to normalize $I(X_k; X_i)$ directly, since $\frac{I(X_k; X_i)}{H(X_i)}$ may not fall in the range $[0, 1]$. Considering $0 \leq I(X; Y) \leq \min\{H(X), H(Y)\}$, we can further define $NI(X_k; X_j) = \frac{I(X_k; X_j)}{\min\{H(X_k), H(X_j)\}} \in [0, 1]$. Therefore, for a target road segment r_i, we iteratively select the feature $r_k \in \mathcal{C}_i$ that maximizes Eq. (4.1), implying that r_k has a large dependency with r_i and small redundancy with current $\mathcal{F}_i$. We repeat this process until κ features are selected or $NJ(X_k) \leq 0$.

(3) ANN based vertex state inference. Once the correlated vertices $\mathcal{F}_i$ are selected, the un-annotated vertex r_i can locally train an ANN model to capture its traffic correlations with the correlated vertices in $\mathcal{F}_i$ using their historical traffic speeds. Specifically, we normalize the historical traffic speeds by comparing with speed limit c^{max} of each vertex respectively during the ANN training. After successfully learning the correlation model, vertex r_i feeds the normalized states of vertices in $\mathcal{F}_i$, extracted from received messages, into the ANN model to infer its own state c_i. Similarly, we normalize the vertex state c_j of $r_j \in \mathcal{F}_i$ as $\frac{c_j}{c_j^{max}}$ using its speed limit, and recover the state of vertex r_i as the product of model inference and speed limit c_i^{max}. The normalization can unify the speed scales of road segments in different road types and transform the input data to the range $[-1, 1]$, where the activation function tanh has the best non-linear transformation capability. Note that vertex r_i only uses the built ANN model once in current time slot and it may need to learn a new model with different correlated vertices in next time slot due to the evolution of property graph. Finally, vertex r_i updates its state as annotated and changes flag as *true*. In the subsequent iterations, r_i will send messages to its neighbors for inferring other unknown vertex states.

4.4 Implementation in Spark

Ideally GPTE should implement the algorithms introduced in previous section into an integrated computing framework that supports both data-parallel and graph-parallel computation. The emerging cluster computing framework Spark [19], which builds on the abstraction *Resilient Distributed Datasets (RDDs)*, supports such requirements. In addition to dataflow operations, the graph processing engine GraphX [5] is built atop of Spark, which represents graph-structured data as a property graph including a pair of vertex RDDs and edge RDDs and embeds graph computations as specific *join-map-group-by* dataflow operators on these RDDs.

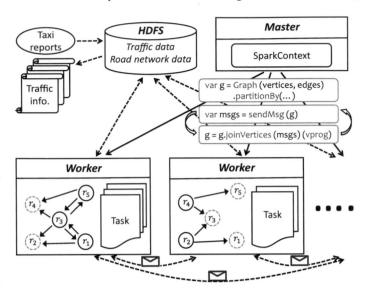

Fig. 4.3 The execution flow of GPTE in a Spark cluster, which consists of one master and some workers. The property graph is distributed to workers for parallel processing. The solid lines represent command flows and dashed lines represent data flow within the cluster

Basic implementation. GPTE is implemented on the Spark and makes use of some GraphX APIs. Figure 4.3 illustrates the execution flow of GPTE in a Spark cluster, which adopts the Hadoop distributed file system (HDFS) for distributed data storage. In the cluster, one machine is selected as the *Master* to coordinate parallel computing among other machines, i.e., *Workers*. GPTE first inputs the road network data from the HDFS to construct the property graph (via `Graph[V, E]()`). Without any optimization, the property graph is distributed to worker machines using the default graph partitioner of GraphX (via `partitionBy`), which exploits a hashing function to evenly distribute vertices and edges among all workers for load balancing. GPTE then continuously conducts traffic estimation from each time slot of traffic samplings. It iteratively lets annotated vertices send messages to neighbors (via `sendMsg`) and then applies the received messages to the property graph (via `joinVertices`), which allows un-annotated vertices to infer their own states (via `vprog`). The vertex program `vprog` embeds correlated vertex selection and ANN modeling based vertex state inference, as mentioned in Sect. 4.3.2.

To improve the system efficiency, we propose (1) a geography-aware partitioner that optimizes the placement of edges/vertices and traffic data among machines to reduce unnecessary communications across different worker machines; and (2) an efficient information propagation scheme including multi-hop message broadcast and redundant message elimination to optimize the information exchange. These techniques can significantly reduce the intra-machine and inter-machine communications and finally improve the system efficiency.

4.4.1 Geography-Aware Graph Partitioning

Due to random edge/vertex assignments of hash-partitioner, the default graph partitioning schemes of GraphX may place neighboring vertices onto different machines, which incurs unnecessary communications and thus results in poor performance. We thus expect a better graph partitioner tailored to the traffic estimation.

The partitioner used in GPTE first maps vertices to real-world geo-coordinates using the middle points of road segments. A simple and effective approach named Hilbert space-filling curve is used to index vertices. Space-filling curve allows one to map multi-dimensional data, e.g., 2-dimensional locations, to 1-dimensional keys that preserve spatial proximity [12]. Keys that are contiguous present nearby locations in space. We thus assign the space-filling curve keys to vertices based on their geographical information. The keys are then range-partitioned into disjoint clusters of nearly equal size. Edges are co-partitioned with vertices by assigning them the same keys as their source vertices. Figure 4.4a illustrates how we divide the road network space into 4×4 cells using Hilbert space-filling curve and assign keys to vertices according to their location coordinates. Based on the keys, we divide the vertices and edges into two partitions. Figure 4.4b shows the pseudocode of key assignment. There are some other techniques, e.g., Quad-tree, can be used to partition the graph while preserving the spatial proximity of vertices. It is, however, difficult for them to evenly control the partition sizes, and thus cannot balance the workloads among different machines, which will finally affect the system performance.

As expressways typically interact with other roads in traffic only at few entries and exits, we thus separately apply the spatial indexing technique to expressway vertices and the other vertices, and then unify the indexes of all vertices (and edges) according to their space-filling keys. We first re-index expressway vertices and then

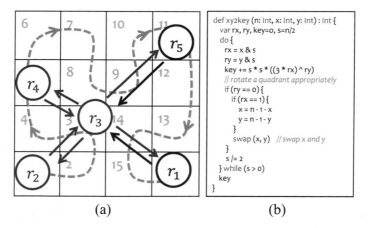

(a)   (b)

Fig. 4.4 **a** Illustration of Hilbert space-filling curve based graph partitioning, where the property graph in Fig. 4.2 is divided into two partitions (in blue and red colors). **b** The pseudocode of key assignment for a given vertex location coordinate (x, y)

```
def sendMsg ( graph: Graph[V, E] ): RDD [Long, Map [Long, Double]] = {
  graph.vertices
      // only annotated vertices send messages
      .filter ( v.flag && v.nblist.length > 0 )
      // propagate messages to all neighbors
      .flatMap ( v.nblist.map ( ( nb, Map ( v.id -> v.state )))
      // aggregate messages targeting to the same vertex
      .reduceByKey ( _ ++ _ )
}
```

Fig. 4.5 Pseudocode of message propagation in Spark

the others, both following the ascending order of their original space-filling keys. Finally, we obtain the continuous vertex keys and then range-partition vertices (and edges) given the desired number of partitions.

4.4.2   Efficient Information Propagation

We propose two techniques to optimize information propagation among vertices. One allows multi-hop message broadcast to speedup the state inferences of all vertices and the other eliminates redundant messages to further reduce communication costs. Figure 4.5 shows the pseudocode that implements message propagation in the Spark.

Multi-hop message broadcast. Since road segments distant from each other may still be traffic correlated, it is necessary for one vertex to propagate its messages to vertices within multiple hops rather than only direct neighbors. To achieve this, a simple approach is based on the hop-by-hop message propagation, where each vertex receives a message and then forwards the message to the next hop until desired hops are reached. Such an approach, however, is not efficient. In GraphX, message propagation is implemented with `triplets` (a triplet contains an edge and its property, and the two vertex properties), which are derived through join operations on vertex RDDs and edge RDDs. As join operation has to be repeated for every hop, the hop-by-hop message propagation introduces excessive computations. We propose multi-hop message broadcast instead, which allows one vertex to broadcast messages to vertices within h hops.

GPTE maintains a `broadcast variable`[4] $\mathcal{B}_{nb}$, which stores the 1-hop neighbors for each vertex. Each vertex r_i in different machine can easily access the variable $\mathcal{B}_{nb}$ and build its h-hop neighbor list $\mathcal{N}_i^h$. For vertex r_i, $\mathcal{N}_i^1$ includes its immediate neighbors, and the h-hop ($h > 1$) neighbor list $\mathcal{N}_i^h$ is recursively built as $\mathcal{N}_i^h = \bigcup_{r_j \in \mathcal{N}_i^{h-1}} \mathcal{N}_j^1$. Based on $\mathcal{N}_i^h$, vertex r_i can directly send messages to its h-hop neighbors via `map` operator. To reduce the communication costs, GPTE first

---

[4]A broadcast variable is a static lookup table and a copy is maintained in each machine to facilitate data access in the Spark.

combines all messages targeting to the same vertex and then sends the aggregated message. Figure 4.5 shows that GPTE realizes the message broadcast through the *map-reduce-by* operation in the Spark. GPTE applies the h-hop messages to the vertices with join operation only once, which completes message propagation with constant overhead instead of being proportional to the hop count.

Eliminating redundant messages. By default, annotated vertices propagate messages to their neighbors and at the same time all vertices receive the messages. As a matter of fact, such messages are only useful for un-annotated vertices for inferring their states but redundant for the annotated vertices. Therefore, it is desired that annotated vertices only send messages to un-annotated vertices so that a large amount of unnecessary message exchanges can be saved.

We propose redundant message elimination for GPTE to avoid unnecessary computation and communication costs. The key idea is that each annotated vertex tracks the statuses of its h-hop neighbors and only sends messages to those un-annotated neighbors in the subsequent iterations. Directly sending vertex status to other vertices in a distributed setting will introduce extra communication overhead, so GPTE implicitly infers vertex status by leveraging the fact that only annotated vertices propagate messages. If an annotated vertex r_i receives a message from vertex r_j, it means that r_j's state is already known and thus r_i's messages to r_j are redundant. Each vertex r_i maintains an un-annotated neighbor list $\overline{\mathcal{N}_i^h}$, which is initially copied from $\mathcal{N}_i^h$ in each time slot. For each annotated vertex r_i, it removes a neighbor r_j from $\overline{\mathcal{N}_i^h}$ once it receives a message from r_j. Similarly, r_i will also be removed from $\overline{\mathcal{N}_j^h}$ by vertex r_j as r_i sends messages to r_j as well. In each iteration, r_i only propagates messages to the neighbors in $\overline{\mathcal{N}_i^h}$. With more and more vertices annotated in later iterations, there are much fewer messages to be processed and thus the system efficiency is improved. To implement this idea in the Spark, GPTE replaces the neighbor list, i.e., v.nblist in Fig. 4.5, as $\overline{\mathcal{N}_i^h}$ and dynamically maintains the list for vertex r_i during the running time.

4.5 Experimental Evaluation

We conduct trace-driven experiments with real-world mobility dataset to evaluate GPTE. We first describe the experiment setup, then evaluate the overall performances of GPTE, and finally run detailed experiments to evaluate the GPTE design.

4.5.1 Experiment Setup

We evaluate GPTE with real mobility data collected by taxis in a cluster.

Road network. The road network covers the center area of our testing city and contains 58356 road segments of different road types, e.g., expressway, major roads, minor roads, and local roads. In the road network, two driving directions of a road are separately represented.

Taxi dataset. The dataset contains traffic samplings collected from more than 12,000 taxis covering the road network of our testing city in July and August of 2015. Each taxi sends back a report every 30 s, which includes a timestamp, GPS location, travel speed, direction, status (*available* or *busy*), etc. About 10 million reports are collected each day. We adopt an accurate map matching algorithm [10] to match each traffic sampling to the road segment a taxi actually traveled, which filters the errant traffic samplings due to GPS noises as well. In addition, we only keep the traffic samplings with taxi status as *busy* for experiments as they reflect more representative conditions of normal traffic.

For evaluations, we keep the traffic data from the last week of August for testing and all the rest as historical data. During the testing phase, we randomly select ρ percents of all taxis in each time slot as *probe taxis* and use their traffic samplings for the traffic estimations. GPTE makes use of the vertex states annotated by $\lambda \geq 5$ probe taxis to infer unknown vertex states on the property graph. We perform traffic estimations for 5:00AM to 21:00PM everyday, which is the typical time period that contains more traffics in the testing city.

Evaluation metrics. During each time slot, we calculate average traffic speed c_i of road segment r_i using all traffic samplings and consider the speed c_i annotated by ≥ 5 taxis as the *ground truth*. The estimation accuracy on r_i is defined as

$$accuracy = \left(1 - \frac{|c_i - \hat{c}_i|}{c_i}\right) \times 100\%, \tag{4.2}$$

where c_i is the ground truth and $\hat{c}_i$ is the estimation. We use the execution time of traffic estimation to measure the computation efficiency.

The Spark configuration. We implement GPTE on the Apache Spark 2.1.0 [1]. For GPTE, we set $h = 5$ to enable each vertex broadcast its messages to neighbors within 5 hops. We set $\kappa = 3$ and thus each vertex selects at most 3 correlated vertices to build the ANN model. These settings are empirically determined in order to balance the estimation accuracy and computation overhead.

Our evaluation environment consists of 5 machines forming a cluster. Each machine has 24 Intel Xeon(R) 3.07 GHz CPU cores and 24 GB of memory. All machines share a disk of size 2 TB. All the road network data and taxi data are stored in the HDFS for distributed data processing. To run GPTE in the Spark, we set one machine as the master machine and treat all the 5 machines as the worker machines, which will perform the ANN modeling and traffic estimations.

4.5.2   Experiment Results

Overall performance. According to our statistics, the number of all taxis (probe taxis) ranges from 1388 to 4942 (972 to 3460) in each time slot throughout the test-

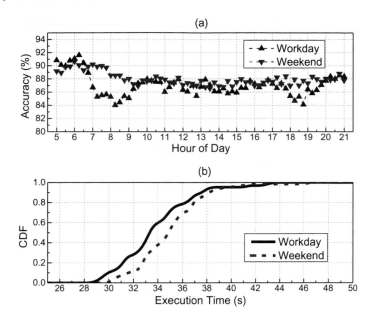

Fig. 4.6 Overall performance on **a** accuracy across the time of a day; **b** execution time

ing week. On average, 17716 road segments (~30.4% of all) are traveled by taxis in a time slot. Figure 4.6a presents the average estimation accuracy across the time of a day. GPTE achieves relatively higher and more stable accuracy on weekends than workdays. This is possibly because the traffic on weekends is more regular than on workdays and thus the traffic correlations are better modeled. The average accuracy for workdays and weekends is 87 and 88%, respectively. We observe obvious accuracy decrease in the time range [6:30AM–9:00AM] and [17:30PM–19:00PM] on workdays, which are the commuter rush hours in Singapore. Surprisingly, GPTE still achieves high accuracy above 83% in these rush hours that always contain heavy and complex traffic.

We plot the statistics of execution time of traffic estimations in Fig. 4.6b. The 90-percentile and 50-percentile execution time are 37 and 33 s for workdays, and 38 and 35 s for weekends, respectively. GPTE runs a bit longer on weekends as there are less taxi data and thus more state inference tasks. The overall average is 34 s, which means that GPTE can return traffic estimation results with a small delay. It is worthy to note that training an ANN model in the HP Z440 Workstation takes about 1.48 s, and it will need much more time to build ANN models for all road segments due to the large computation overhead, e.g., tens of minutes even we adopt the multi-threading technique, which is much larger than the time budget of 15 min. Therefore, Fig. 4.6b demonstrates that GPTE can significantly improve the computation efficiency with high parallelism.

Next we will show how the proposed optimization techniques help GPTE achieve timely and accurate traffic estimations.

Table 4.1 The cross-partition messages for different graph partitioning schemes

Scheme	Space-filling	Quad-tree	Random	EP1D	EP2D
Average $(\times 10^3)$	11.8	31.3	145.1	144.9	145.2

Impact of geography-aware graph partitioning. To understand the impact of data placement on system efficiency, we compare our geography-aware graph partitioner using space-filling curve technique with the Quad-tree based graph partitioner and three default graph partitioning schemes of GraphX, i.e., *RandomVertexCut* (Random), *EdgePartition1D* (EP1D), and *EdgePartition2D* (EP2D). The Quad-tree based method recursively subdivides a graph into four quadrants until the number of vertices in each quadrant is smaller than a threshold. This method preserves spatial proximity of vertices as well but it unevenly distributes vertices among partitions. The default graph partitioner assigns edges/vertices to different partitions by hashing source (or destination) vertex ID, resulting in random placements of edges/vertices among machines. In contrast, our scheme exploits vertex locations to place nearby edges/vertices to the same machine.

To compare the communication efficiency of different schemes, we count the number of cross-partition messages (communication overhead). During each traffic estimation, about 153.3×10^3 messages in total are generated for information propagation among vertices, while there are on average $11.8 \times 10^3, 31.3 \times 10^3, 145.1 \times 10^3$, $144.9 \times 10^3, 145.2 \times 10^3$ cross-partition messages for the five schemes respectively, as shown in Table 4.1. From the statistic, we see that space-filling method significantly outperforms other methods, with reduction on cross-partition messages by $4.4\times$ than Quad-tree and $11.4\times$ than the three default schemes of GraphX.

In addition, we use execution time as another metric to compare these five schemes since cross-partition messages could be costly and will explicitly affect the execution time. The comparison results are depicted in Fig. 4.7a. Compared to default schemes of GraphX, the benefit of exploiting vertex locations is clear that the proposed scheme reduces the execution time up to 6 s. When vertices that are geographically close to each other are assigned to the same partition, the data exchanges among different partitions are reduced. Although Quad-tree based scheme has also exploited geographical information of vertices, it cannot evenly assign vertices to partitions that lead to unbalanced loads among machines. Relying on the Hilbert space-filling curve technique, our scheme indexes vertices with consecutive keys and can evenly partition vertices among machines, which is more flexible and efficient. Besides, our scheme makes graph processing more stable when compared with other schemes on the variance of execution times.

Impact of efficient information propagation. Rather than propagating messages among vertices in a hop-by-hop manner, we propose the multi-hop message broadcast to accelerate information propagation. Indeed we can first learn h-hop neighbors for each vertex and then build another graph where vertices are still the road segments and edges are formed between each vertex and its h-hop neighbors. In this graph,

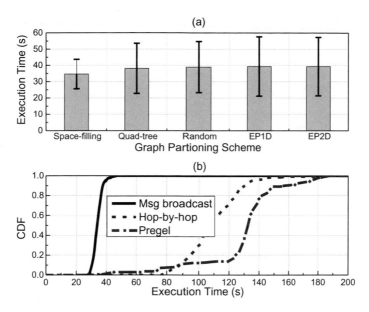

Fig. 4.7 Evaluation of **a** graph partitioning; **b** message broadcast

each vertex can directly send messages to its original h-hop neighbors through the `triplets` in GraphX. We implement this idea using `Pregel` [9] API of GraphX. We compare the three message propagation methods and plot the results in Fig. 4.7b. Average execution times of multi-hop message broadcast, hop-by-hop method, and Pregel based method are 34, 110, and 130 s, respectively. Our method significantly outperforms the hop-by-hop method and Pregel based method with gains of 3.2× and 3.8×, respectively. The multi-hop message broadcast avoids repetitive join-operations between vertex RDDs and edge RDDs when compared to the hop-by-hop method, and meanwhile keeps the original road network structure rather than generating a particular graph as Pregel based method dose.

In addition, we propose redundant message elimination to further improve the efficiency of information propagation. By avoiding redundant messages to anno-tated vertices, GPTE can reduce communication costs. Figure 4.8 shows the execu-tion details of one sample traffic estimation. This job lasts for 27 iterations and we report the execution time and number of messages in each iteration for the scenarios *with* and *without* redundant message elimination. Without such a mechanism, there are more and more message exchanges as more vertex states are inferred along with the time. Execution time per iteration maintains for about 3 s. In contrast, eliminat-ing redundant message significantly reduces the number of messages for the later iterations and execution time of each iteration is continuously decreased as shown in Fig. 4.8. Overall, redundant message elimination reduces the average execution time from 101 to 34 s, providing gains of ∼3×.

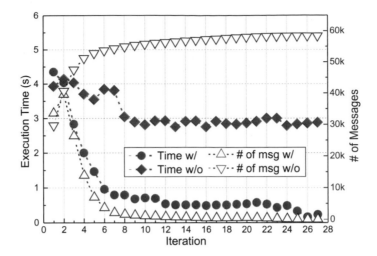

Fig. 4.8 The execution details of one sample traffic estimation

4.6 Summary

This chapter presents GPTE for accurate and timely traffic estimations at city scale. GPTE proposes non-linear traffic correlation modeling and adapts the advanced traffic estimations to the graph-parallel processing paradigm through dynamic ANN modeling and efficient information propagation. A series of optimization techniques are proposed to improve the design. Experiment results from real-world mobility data demonstrate that GPTE can derive traffic monitoring for the entire city in 34 s with accuracy as high as 88%.

References

1. Apache Spark. http://spark.apache.org/
2. Asif, M.T., Dauwels, J., Goh, C.Y., Oran, A., Fathi, E., Xu, M., Dhanya, M.M., Mitrovic, N., Jaillet, P.: Spatiotemporal patterns in large-scale traffic speed prediction. IEEE Trans. Intell. Transp. Syst. **15**(2), 794–804 (2014)
3. Aslam, J., Lim, S., Pan, X., Rus, D.: City-scale traffic estimation from a roving sensor network. In: ACM SenSys (2012)
4. Estévez, P.A., Tesmer, M., Perez, C.A., Zurada, J.M.: Normalized mutual information feature selection. IEEE Trans. Neural Netw. **20**(2), 189–201 (2009)
5. Gonzalez, J.E., Xin, R.S., Dave, A., Crankshaw, D., Franklin, M.J., Stoica, I.: GraphX: graph processing in a distributed dataflow framework. In: USENIX OSDI (2014)
6. Liu, Z., Li, Z., Li, M., Xing, W., Lu, D.: Mining road network correlation for traffic estimation via compressive sensing. IEEE Trans. Intell. Transp. Syst. **17**(7), 1880–1893 (2016)
7. Liu, Z., Zhou, P., Li, Z., Li, M.: Think like a graph: real-time traffic estimation at city-scale. IEEE Trans. Mob. Comput. **18**(10), 2446–2459 (2018)

8. Lv, Y., Duan, Y., Kang, W., Li, Z., Wang, F.-Y.: Traffic flow prediction with big data: a deep learning approach. IEEE Trans. Intell. Transp. Syst. **16**(2), 865–873 (2015)
9. Malewicz, G., Austern, M.H., Bik, A.J., Dehnert, J.C., Horn, I., Leiser, N., Czajkowski, G.: Pregel: a system for large-scale graph processing. In: ACM SIGMOD (2010)
10. Newson, P., Krumm, J.: Hidden Markov map matching through noise and sparseness. In: ACM SIGSPATIAL (2009)
11. Peng, H., Long, F., Ding, C.: Feature selection based on mutual information criteria of max-dependency, max-relevance, and min-redundancy. IEEE Trans. Pattern Anal. Mach. Intell. **27**(8), 1226–1238 (2005)
12. Sagan, H.: Hilbert's space-filling curve. Space-Filling Curve, pp. 9–30. Springer, New York (1994)
13. Tan, H., Feng, G., Feng, J., Wang, W., Zhang, Y.-J., Li, F.: A tensor-based method for missing traffic data completion. Transp. Res. Part C: Emerg. Technol. **28**, 15–27 (2013)
14. Vlahogianni, E.I., Karlaftis, M.G., Golias, J.C.: Short-term traffic forecasting: where we are and where we're going. Transp. Res. Part C: Emerg. Technol. **43**, 3–19 (2014)
15. Wang, Y., Zheng, Y., Xue, Y.: Travel time estimation of a path using sparse trajectories. In: ACM SIGKDD (2014)
16. Xin, X., Lu, C., Wang, Y., Huang, H.: Forecasting collector road speeds under high percentage of missing data. In: AAAI (2015)
17. Yang, B., Guo, C., Jensen, C.S.: Travel cost inference from sparse, spatio temporally correlated time series using Markov models. Proc. VLDB Endow. **6**(9), 769–780 (2013)
18. Yang, B., Kaul, M., Jensen, C.S.: Using incomplete information for complete weight annotation of road networks. IEEE Trans. Knowl. Data Eng. **26**(5), 1267–1279 (2014)
19. Zaharia, M., Chowdhury, M., Das, T., Dave, A., Ma, J., McCauley, M., Franklin, M.J., Shenker, S., Stoica, I.: Resilient distributed datasets: a fault-tolerant abstraction for in-memory cluster computing. In: USENIX NSDI (2012)
20. Zheng, J., Ni, L.M.: Time-dependent trajectory regression on road networks via multi-task learning. In: AAAI (2013)
21. Zhu, Y., Li, Z., Zhu, H., Li, M., Zhang, Q.: A compressive sensing approach to urban traffic estimation with probe vehicles. IEEE Trans. Mob. Comput. **12**(11), 2289–2302 (2012)

Chapter 5
A Crowdsensing Based Traffic Monitoring Approach

Abstract This chapter presents a crowdsensing based urban traffic monitoring system. Different from existing works that heavily rely on intrusive sensing or full cooperation from probe vehicles, our system exploits the power of participatory sensing and crowdsources the traffic sensing tasks to bus riders' mobile phones. The bus riders are information source providers and meanwhile major consumers of the final traffic output. The system takes public buses as dummy probes to detect road traffic conditions, and collects minimum set of cellular data together with some lightweight sensing hints from the bus riders' mobile phones. Based on the crowdsourced data from participants, the system recovers the bus travel information and further derives the instant traffic conditions of roads covered by bus routes. The real-world experiments with a prototype implementation demonstrate the feasibility of our system, which achieves accurate and fine-grained traffic estimations with modest sensing and computation overhead at the crowd.

Keywords Crowdsensing · Bus systems · Bus riders · Cellular signal · Trajectory mapping

5.1 Introduction

Recently people take[1] the roving vehicles, e.g., taxis, on roads as probes to detect the instant traffic conditions [10, 12]. Although these passive probing methods can extract on-site information from probes freely roving in the city with lightweight cost, they still cannot provide complete traffic estimations for the whole road network due to insufficient probes. In addition, as probe vehicles are usually managed by transit companies or agencies, it requires substantial efforts for obtaining the mobility data.

In this chapter, we present a crowdsensing based urban traffic monitoring system, which takes public buses as probes to sample the instant road traffic conditions. In urban cities, the public buses cover most roads with a high coverage ratio of the

---

[1]Parts of this chapter is reprinted from [6], with permission from IEEE.

© The Author(s), under exclusive license to Springer Nature Singapore Pte Ltd. 2021   49
Z. Liu and K. Wu, *Mobility Data-Driven Urban Traffic Monitoring*,
SpringerBriefs in Computer Science,
https://doi.org/10.1007/978-981-16-2241-0_5

whole road network. Instead of requesting the GPS traces from any third parity, our system relies on the help of bus riders and crowdsources the traffic sensing tasks to their commodity mobile phones. The mobile phones automatically collect real-time traffic sensing data and anonymously upload the data to a backend server, which is responsible for processing and analyzing the uploaded information from different buses. Bus travel information are extracted and general travel speeds at different road segments are estimated to generate the traffic map of roads covered by public buses. Our system is fully built on the bus riders' commodity mobile phones with low computation and energy costs, which can encourage a wide participation for large service coverage. The lightweight design allows the immediate adoption of our system to other cities.

Despite the advantages, the realization of such a participatory urban traffic monitoring system encounters a set of challenges which call for practical and effective solutions to cope with. First, accurately and efficiently tracking bus trip is non-trivial. Considering the practical requirements of a crowdsensing system, energy-hungry GPS sensor is undesirable. Instead, we prefer the cellular signal together with several lightweight sensing hints, e.g., audio and acceleration signals, from the mobile phones to detect and identify the bus trip information. However, the cellular signals only provide rough location references which are insufficient for precise vehicle tracking [5]. By leveraging the fact that public buses travel along determined routes and stop at known bus stations, we propose a novel method that exploits the invariant locations and cellular attributes of bus stations to build a location mapping between the physical space and the cellular space. Second, the sensing data collected from bus riders are complicated and noisy even with errors. To guarantee accurate traffic estimation, we clean the sensing data at individual mobile phone and propose some clustering and aggregation methods at the backend server to process and analyze the joint data from all participants.

We detail and integrate all above techniques for a complete urban traffic monitoring system, and implement a prototype system on the Android platform and a laboratory server. To evaluate our system, we conduct experiments with 8 bus routes in a $\sim$28 km^2 region. During the 2-month experiments, our system has received data from 122 participants and derived instant traffic map of the testing area. Experimental results demonstrate the feasibility and effectiveness of our system in practice. The system overhead is also carefully investigated.

5.2   Motivation

Although there exists many works about traffic monitoring, the intrusive sensing approaches usually incur huge infrastructure costs and the probe vehicle based approaches are usually limited by data availability. To get rid of these disadvantages, we present a participatory sensing based urban traffic monitoring system that resorts to public buses for probing real-time traffic conditions. Specifically, we fundamentally decompose the traffic sensing tasks from the running buses to bus riders' mobile phones. The bus riders themselves contribute the primary traffic sensing data and meanwhile consume the final traffic output. Such a low-cost and flexible system can be easily adopted in other cities with slight modification. The public bus network covers most of the roads in a city to provide convenient commuting for citizens, which thus provides good coverage of traffic monitoring.

The straightforward approach for vehicle tracking is using GPS sensors [8], however, it may not be a good choice due to the considerations of energy consumption and localization accuracy. First, GPS device is energy aggressive, which sorely discourages user participation due to the limited battery capacity of commodity mobile phones. Second, GPS suffers from large localization error in the downtown area due to the complicated surroundings. It is even worse when the mobile phones are placed inside buses where the GPS signal is further attenuated.

Compared to energy-hungry GPS sensors, cellular signal from mobile phones is more energy-friendly and widely available [4], which makes it a better sensing hint for vehicle tracking. Meanwhile cellular signal outperforms other possible wireless signals, e.g., WiFi [9], for location references due to following advantages. First, mobile phones always keep the cellular module working to support persistent telecommunication services. Thus the marginal energy consumption of collecting cellular signals is negligible. Much extra energy, however, will be consumed for scanning other wireless signals like WiFi [8]. Second, cell towers are widely deployed to provide the complete coverage of the entire city while other wireless signals are usually sporadically available with poor coverage in outdoor areas. Third, unlike other transient wireless signal sources, e.g., WiFi hotspots, cellular signal sources are much more consistent over time, which makes the cellular signature database more stable and easier to maintain.

The typical coverage of a cell tower in the urban area is about $200 \sim 900 \, m^2$. Thus the cellular signals provide only rough location references and are insufficient for instant and accurate bus tracking. The fact that public buses strictly follow the determined bus routes and stop at the known bus stations, however, provides us an opportunity to relax the requirement of precise bus tracking. Thus we transform the precise bus tracking problem to the bus status detection and bus stop identification among all possible bus stops. Based on the bus status information, we can recover the bus movements and estimate the traffic conditions on the road segments in between bus stops. The final traffic map can be derived by assembling the traffic estimations of all covered road segments. To accurately identify bus stops, we need to collect the cellular signals for all bus stops as cellular fingerprints, and later use these fingerprints to match bus stops in cellular space with the cellular signals uploaded from bus riders' mobile phones.

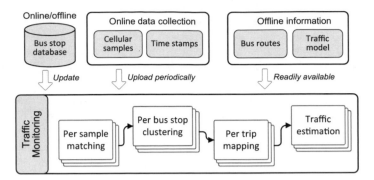

Fig. 5.1 The architecture of our proposed crowdsensing based urban traffic monitoring system

5.3 System Design

The system consists of two major components, i.e., online/offline data collection and trajectory mapping for traffic estimation, as sketched in Fig. 5.1. In the following subsections, we will elaborate each component in details.

5.3.1 Data Collection

As depicted in Fig. 5.1 (top), the system input mainly comes from the following three data sources.

Bus riders. The bus riders serve as the traffic probes along bus routes and are the major information sources of our system. Once the users are detected on buses, our system will automatically start the online data collection from mobile phones. A beep detection approach similar with that in [11] is applied to detect whether the user is on a public bus or not. Nowadays IC card systems are worldwide adopted by bus operators, e.g., ORCA in Seattle, EZ-link in Singapore, Oyster in London, and MetroCard in New York, to automatically collect transit fees. Typically the bus riders tap their IC cards on the card readers to pay their fees when getting on or off the bus at some bus stop, and meanwhile the card readers generate a unique beep, which is always consist of audio signals in specific frequencies, e.g., a combination of 1 and 3 kHz in Singapore and 2.4 kHz in London. To efficiently detect such a beep signal, we prefer the Goertzel algorithm [1] instead of Fast Fourier Transform (FFT) used in [11] to extract specific frequencies rather than all frequencies based on the prior knowledge of frequency components in the beep. The Goertzel algorithm performs tone detection using much less CPU computation than FFT, and thus significantly saves energy. To detect the beeps, we measure and normalize the signal strength of several interested frequency bands. If the signal strength of the frequency bands obviously jumps and exceeds an empirical threshold (e.g., three standard deviation of the signal strength), the beep detection is confirmed. To enhance the robustness of beep detection, we further adopt a standard sliding window averaging method with window size 300 ms to filter out possible noises. The "sound" hints are widely adopted in IC card systems to let passengers be aware of the transactions. Thus our system can work in other cities with presetting the frequency components in audio signal. Even without such "sound" hints, our system can still detect the on-bus status using some inertial sensors based transportation mode detection methods [3].

The mobile phone will automatically start recording a bus trip after a confirmed beep detection. For each thereafter detected beep event, the mobile phone records the set of visible cell tower signals with an attached timestamp. Thus the sensing data on the mobile phone corresponds to a sequence of timestamped cellular samples along the bus trip. These data are uploaded to the backend server anonymously and periodically. The mobile phone terminates current trip if no beep is detected for $\Delta = 10$ mins, which implies the user has got off the bus. Once new beeps are thereafter detected, the mobile phone will start recording and uploading another independent bus trip. The parameter Δ can be adjusted for different cities. To detect possible traffic jams between two consecutive bus stops where bus travel time exceeds Δ, our system tracks the trip status of each anonymous mobile phone. Specifically, if two consecutive cellular samples are collected from the same bus route, we treat they belong to the same trip; otherwise, we think the user starts a new trip.

For the beep detection, we continuously filter out the noisy beep detection (e.g., the beeps in rapid train stations using the same IC card systems or the beeps false positively detected by the users waiting for other buses at the bus stops) by analyzing the readings of accelerometer on mobile phones to distinguish human mobility. A simple threshold based method is enough to filter out the noisy beeps. In general, buses usually travel with frequent acceleration changes while rapid trains are operated more smoothly. Similarly, the readings collected when the user is walking or standing at the bus stop are much smoother than those collected on a moving bus.

Bus stop database. We assume there is an offline built database which stores the cellular fingerprints for all bus stops. The backend server relies on these cellular fingerprints to identify the corresponding bus stops for each uploaded cellular sample. The bus stop database can also be constructed in an online manner, and the readers are encouraged to find more details in [6].

Bus routes and traffic model. The bus operators publicly publish the bus operational route information on the web, which implies constraints on how bus stops are passed. We make use of such information for trajectory mapping later. In addition, some mature and classical traffic models [2], which describe the transformation between travel speeds of buses and general vehicles, are available for us to derive the general traffic conditions from buses. Therefore, although we reply on public buses to probe traffic conditions, our system reports the general travel speeds that are useful for all vehicles.

5.3.2 Trajectory Mapping

As depicted in Fig. 5.1 (bottom), we take the bus stops as landmarks and match the uploaded cellular samples to map the bus trajectories. For the received sequence of cellular samples from each independent trip, the backend server will recover the bus trip information by identifying the passing by bus stops. The server will do three levels of mapping to enhance the accuracy of trajectory mapping.

5.3.2.1 Per Sample Matching

For a typical IC card system, the card readers are only enabled when the buses arrive at the bus stops, where bus riders pay their transit fees by tapping IC cards. The beep, detected by bus riders' mobile phones, thus indicates the bus arrival at a bus stop, and the collected cellular sample at that time corresponds to a particular bus stop.

To identify one bus stop for each cellular sample of one trip, we match it with the signature sets stored in the fingerprint database. We adopt the modified Smith-Waterman algorithm [11] to measure the similarity of different cell ID sets. This algorithm focuses on the orders rather than the absolute RSS values of cell towers, which thus tolerates possible variances of RSS values due to different conditions (e.g., on/off buses, weather, time, etc.). The backend server reorganizes the cell IDs in a set in a descending order according to their RSS values. For the cell ID set of each cellular sample, the algorithm compares all possible segments to determine the optimal alignment with one cellular signature in the fingerprint database, and assigns different weighs to the matching results of *match*, *mismatch* and *gap*. As the penalty cost for gaps and mismatches will significantly affect the matching performance, we conduct simulations to select the best penalty cost by varying its value from -0.1 to -0.9. Simulation shows that -0.3 as the penalty cost achieves the best matching accuracy. We illustrate this algorithm with an example in Table 5.1 where the uploaded cellular sample with cell ID set as $c_{up} = \langle 1, 2, 3, 4, 5 \rangle$ is compared with one cellular fingerprint $c_{db} = \langle 1, 7, 3, 5 \rangle$. The algorithm finally scores 2.4 by aggregating 3 matches, 1 gap and 1 mismatch.

The matching algorithm runs over all bus stop candidates in the database, and selects one bus stop with the highest similarity score. To guarantee the accuracy of trajectory mapping, we further filter out possible noise cellular samples whose highest matching scores with candidate bus stops are lower than a threshold. We empirically

Table 5.1 Bus stop matching instance for one cellular sample which contains 5 cell tower IDs

c_{up}	1	2	3	4	5	Match	Gap	Mismatch	$\sum$
c_{db}	1	7	3		5	3	1	1	2.4
Match	**1**	×	**3**	–	**5**				
Score	+1	−0.3	+1	−0.3	+1				

set the threshold as 2. Thus all cellular samples with low highest similarity scores are discarded with no further processing. If more than one bus stops are matched with one cellular sample, we select the one with a larger number of common cell IDs. Such a tie-breaker setting can effectively determine one bus stop for each valid cellular sample according to our practical tests. Besides, if the cellular sample $e(x)$ finally matches the bus stop fingerprint $b(y)$ we denote $M(e(x), b(y)) = 1$, otherwise $M(e(x), b(y)) = 0$.

5.3.2.2 Per Bus Stop Clustering

In general, a number of passengers board and alight at a bus stop when the bus arrives, which triggers multiple beeps. As a result, the mobile phones of bus riders can collect and upload multiple cellular samples at one bus stop to the backend server. Such redundant information allows us to further improve the bus stop identification accuracy. We can group the cellular samples according to their matched bus stops and timestamps, and then identify the bus stop for each closely clustered cellular samples with more confidence.

Given a sequence $E = \{e_1, e_2, \ldots, e_m\}$ of m cellular samples with attached timestamps $T = \{t_1, t_2, \ldots, t_m\}$, we can obtain their corresponding bus stops $\{b_1, b_2, \ldots, b_m\}$ with similarity scores $\{s_1, s_2, \ldots, s_m\}$. We cluster cellular samples in the space enabled by three dimensions of time, bus stop, and matching score. In the co-clustering algorithm, for cellular samples corresponding to the same bus stop, we denote the maximum mutual similarity score as s_0 and the maximum possible time interval between their timestamps as t_0, which are empirically set as 7 and 30 s in our system, respectively. For any two cellular samples e_i and e_j, we weigh their matching relationship as

$$L(e_i, e_j) = \begin{cases} \frac{s_0 - |s_j - s_i|}{s_0}, & \text{if } b_i = b_j \\ 0, & \text{otherwise.} \end{cases}$$

Considering the timestamp information, we put e_i and e_j into the same cluster if

$$\frac{t_0 - |t_j - t_i|}{t_0} + L(e_i, e_j) > \varepsilon, \tag{5.1}$$

where ε is a threshold to verdict whether e_i and e_j belong to the same cluster. Only when two cellular samples are collected close in time and have approximate similarity scores, they will be grouped into the same cluster. We choose $\varepsilon = 0.6$ in the system implementation.

By applying the co-clustering on all cellular samples, we finally derive a sequence of n clusters $\{C_1, C_2, \ldots, C_n\}$. Each cluster C_i should correspond to one bus stop. Due to the noises, however, the cellular samples from the same cluster may match different bus stops. To guarantee the accuracy of bus stop identification, each cluster C_i is thus temporarily associated with several potential bus stop candidates, as

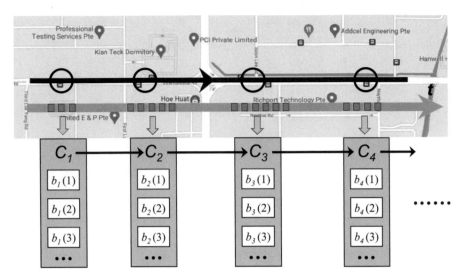

Fig. 5.2 Bus stop identification with a sequence of clusters. Each cluster contains multiple bus stop candidates

demonstrated in Fig. 5.2. In practice, however, most clusters only have one bus stop candidate according to our experiments.

5.3.2.3 Per Trip Mapping

The pre-configured bus routes greatly constrain the possible sequences or combinations of bus stops that can be visited by a specific bus. Such constraints help us filter out the impossible bus stop candidates and finally match each cluster of cellular samples to a sole bus stop. Figure 5.2 depicts a concrete example where a sequence of n clusters are derived from all cellular samples. Each cluster $C_k (k = 1, 2, \ldots, n)$ includes E_k cellular samples $\{e_k(1), e_k(2), \ldots, e_k(E_k)\}$ and B_k bus stop candidates $\{b_k(1), b_k(2), \ldots, b_k(B_k)\}$. We assign each bus stop candidate $b_k(i)$ a probability $p_k(i) = \frac{\sum_{j=1}^{E_k} M(e_k(j), b_k(i))}{E_k}$ and an average similarity $\bar{s}_k(i) = \frac{\sum_{j=1}^{E_k} [M(e_k(j), b_k(i)) \cdot Sim(e_k(j), b_k(i))]}{\sum_{j=1}^{E_k} M(e_k(j), b_k(i))}$. Now we try to find out a segment from one bus route or the possible concatenation of multiple different bus routes that best matches the current trip and then derive the most "correct" bus stop for each cellular sample cluster.

For any two bus stops x and y, we denote their order relationship in bus routes as $R(x, y) = 1$ if y is later visited by buses after passing by x in some bus route, $R(x, y) = 0$ if $x = y$, and $R(x, y) = -1$ for the rest. As some cellular sample cluster may match more than one bus stops, we thus derive a set of bus stop sequences $S = \{S_1, S_2, \ldots, S_N\}$, where $N = \prod_{k=1}^{n} B_k$. Each S_j represents a sequence of n bus stops as $\{b_1(S_j(1)), b_2(S_j(2)), \ldots, b_n(S_j(n))\}$. We determine the bus stop sequence best matched with current trip using the maximum likelihood estimation as

$$
\begin{aligned}
S^* = \arg \max_{S_{j:1\sim N}} \{ p_1(S_j(1)) \cdot \bar{s}_1(S_j(1)) + \sum_{i=2}^{n} [p_i(S_j(i)) \\
\cdot \bar{s}_i(S_j(i)) \cdot R(b_{i-1}(S_j(i-1)), b_i(S_j(i)))] \},
\end{aligned}
\tag{5.2}
$$

where we weigh S_j using both the probabilities $p_i(S_j(i))$ and average similarities $\bar{s}_i(S_j(i))$. The output S^* finally recovers the trajectory of current trip in the form of the best matched bus stop sequence, which also determines the most likely bus stop for each cellular sample cluster on the trajectory.

5.3.3 Traffic Estimation

In this subsection, we describe how we make use of trajectory mapping results to estimate traffic conditions of road segments in between bus stops on the trajectories.

Based on the matched bus stop sequences, we first purge the cellular samples by filtering the noise samples which falsely match other bus stops. Then by ordering the cellular samples in the same cluster according to their timestamps, we can extract the arrival time and departing time of one bus at the corresponding bus stop. Figure 5.3 illustrates an example with a set of cellular samples from the same trip. These cellular samples are classified into 2 clusters corresponding to two bus stops, i.e., i and j. Based on the cellular sample cluster for bus stop i, we can extract the corresponding arrival time $t_a(i)$ and departing time $t_d(i)$. The same information can be derived

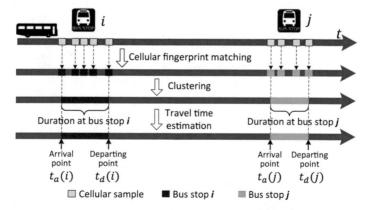

Fig. 5.3 Bus stop clustering and time duration at bus stops of an example trip. 5 samples are collected at stop i and 4 samples are collected at stop j, which are clustered into 2 groups. The bus arrival time and departing time at bus stops are estimated and used for travel time estimation

for bus stop j. Then we estimate the bus travel time between stop i and j as $t_{ij} = t_a(j) - t_d(i)$. In practice, some bus stops may be skipped by the buses if no passengers board or alight, which results in information missing at these bus stops. In such cases, our system automatically treats the adjacent road segments as one and estimate the travel time on the combined road segment.

The travel time on a road segment by public buses may not accurately reflect the practical traffic condition. There exists a gap between the travel time of general automobiles ($\mathcal{T}_A$) and that of buses ($\mathcal{T}_B$) on the same road segment. The difference arises mainly due to the special operations of buses, e.g., frequent stopping at bus stops for passengers boarding and alighting, repetitive accelerations and decelerations from and to bus stops. Besides, there is a natural difference between buses and general automobiles due to their different operating abilities and speed limits. Fortunately, the relationship between the two kinds of travel times has been well studied in the transportation domain [2, 7]. We have tested different models, both linear and non-linear models. The experiment results suggest that compared to non-linear models, the linear models are good enough for the travel time transformation, which are easy to learn without complex parameter settings and over-fitting issues. Thus we use one classical linear traffic model proposed in [2] to estimate $\mathcal{T}_A$ from $\mathcal{T}_B$:

$$\mathcal{T}_A = a + b \times \mathcal{T}_B, \tag{5.3}$$

where $a = \frac{road\ length}{free\ travel\ speed}$ represents the average travel time by a typical automobile on the road segment when there is little or no traffic, and b represents the traffic congestion effect on $\mathcal{T}_A$, as measured by the travel time of public buses. Based on historical traffic data, we can learn the parameter b using ordinary least-square technique. According to our experiments, we find that the best values of b for different road segments fall in a narrow range $[0.13, 0.18]$. In our system, we set $b = 0.15$ for all road segments for simplicity. The average travel speed of general automobiles on the road segment can thus be estimated as $v_A = \frac{road\ length}{\mathcal{T}_A}$. Although some cities may design dedicated lanes for a few special buses, e.g., bus rapid transit, there still are many ordinary buses traveling on the roads. To guarantee the correctness of our estimations, our system can filter out the trip reports from the special buses by comparing the mapped bus trajectory with special bus routes.

For each road segment, it may be simultaneously covered by multiple bus routes, which leads to several speed estimations as our system collects trip reports from massive mobile phones on various buses. Thus our system adopts a Bayesian method [7] to continuously update the traffic estimation by carefully combining previous estimation and current estimation from new data input. With the variance of previous average speed $\bar{s}_0$ as σ_0^2 and the variance of new average speed $\bar{s}$ as σ^2, the updated speed estimation is normal with average speed $\bar{s}_{new}$ and variance σ_{new}^2 as

$$\bar{s}_{new} = \frac{\frac{\bar{s}_0}{\sigma_0^2} + \frac{\bar{s}}{\sigma^2}}{\frac{1}{\sigma_0^2} + \frac{1}{\sigma^2}}, \quad \sigma_{new}^2 = \frac{1}{\frac{1}{\sigma_0^2} + \frac{1}{\sigma^2}}, \tag{5.4}$$

which uses the inverse of estimation variance to weigh previous estimation and new estimation. The updating procedure produces sequential travel speed estimations with newly received traffic samplings from bus riders.

We combine the traffic estimations of all road segments to derive a complete traffic map. In particular we weight the overlapping road segments in combining their estimations. Say two road segments, AC and BC, share the common part IC where I is the intersection point of the two segments. When combining the traffic conditions of AC and BC, we divide them into AI, BI and IC. We weight $\bar{s}_{AC}$ and $\bar{s}_{BC}$ based on the position of I to derive the speed estimation $\bar{s}_{IC}$, i.e.,

$$\bar{s}_{IC} = \frac{\alpha \times \bar{s}_{AC} + \beta \times \bar{s}_{BC}}{\alpha + \beta},$$

where $\alpha = \frac{d_{IC}}{d_{AC}}$, $\beta = \frac{d_{IC}}{d_{BC}}$ and d_{ij} is the road length between i and j. Meanwhile the travel speed on AI and BI, $\bar{s}_{AI}$ and $\bar{s}_{BI}$, can be calculated as

$$\bar{s}_{AI} = \frac{d_{AI}}{\frac{d_{AC}}{\bar{s}_{AC}} - \frac{d_{IC}}{\bar{s}_{IC}}}, \quad \bar{s}_{BI} = \frac{d_{BI}}{\frac{d_{BC}}{\bar{s}_{BC}} - \frac{d_{IC}}{\bar{s}_{IC}}}.$$

The system will update the travel speed estimations on all road segments with a period of $T = 15\,$mins, which is a sufficiently fine-grained and adopted by many previous works for traffic estimations [10].

5.4 Experimental Evaluation

In this section, we evaluate the performance of our system based on a prototype implementation on Android platform. We first introduce the implementation details and experimental settings. Then we present the evaluation results of traffic estimations. Finally, we investigate system overhead.

5.4.1 Experiment Setup

We have implemented the bus trip data collection App on Android platform (with Android version 4.0.3). Controlled experiments are conducted with three types of mobile phones, i.e., HTC Sensation XE, HTC Desire S, and Google Nexus One. These mobile phones are common phones equipped with accelerometer sensors and support 16-bit 44.1 kHz audio signal sampling from microphones. Their memory and CPU capacity are powerful enough for the light computation and bus trip sensing involved in our application. The phone types of the participants are more diverse. The HTC and Samsung phones dominate. As our system requires no particular hardware,

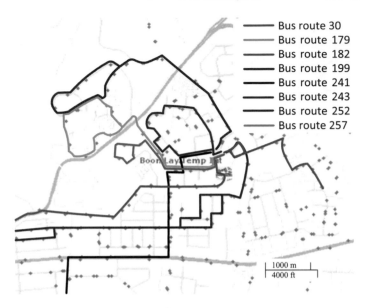

Fig. 5.4 The eight bus routes in the ∼28 km² testing area for our experiments

thus the proposed method could be easily implemented on other OS and hardware platforms, e.g., Apple iPhones and Windows Phones. We also implement the backend bus trip data processing and analysis services in Java executing on a server, i.e., the DELL Precision Workstation T3500, for our experiments.

Figure 5.4 demonstrates our experiment area with a size of about 7 km × 4 km, which owns more than 20 bus routes with periodically transit bus services covering most of the roads in this area. Our experiments concern on 8 bus routes, including bus route 179, 182, 199, 241, 243, 252, 257 and partial part of route 30, as shown in Fig. 5.4. The selected bus routes cover a great portion of the roads in the area and thus can provide fine-grained traffic estimation results. In the testing area, we conduct various experiments to evaluate our system and methods. All the experiments lasted for more than 2 months.

Data collection. The input of our system includes the cellular fingerprints of bus stops and the real-time bus trip sensing data collected from mobile phones of bus riders. We have 122 participants in total involved in our experiments to contribute their real-time bus trip information to our system. The participants installed our data collection App on their mobile phones to collect bus trip data and periodically upload the sensory data through WiFi or 3G/4G to our backend server.

5.4.2   Traffic Estimation Performance

First, we evaluate our traffic estimation method with sensory data collected from participants' mobile phones. Although we rely on traffic data collected from buses,

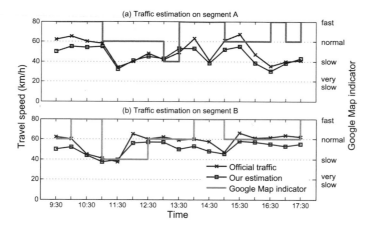

Fig. 5.5 Traffic estimation compared with official traffic data and Google Maps' indicator

we have transformed the bus travel information to the general travel speeds according to the method in Sect. 5.3.3. The following analysis is based on the finally transformed travel speeds.

We compare our estimated traffic speeds with official traffic speeds acquired from the transport authority. The authority accumulates traffic data from traffic samplings of more than 10,000 roving taxis and other data sources including traffic cameras and inductive loop detectors within a time slot of 15 mins and derives the real-time average traffic speeds. The obtained official traffic speeds fully cover the experiment days and the testing area. In addition, we compare our results with Google Maps' traffic data. Google Maps provide the live traffic visualization on the maps but they only give 4 coarse traffic levels (i.e., *very slow*, *slow*, *normal*, and *fast*) instead of providing detailed road traffic speed to the end users. We manually obtain the coarse congestion indicator data from Google Maps that cover the experiment days and testing roads, and then compare them with our estimation results. We pick 2 typical road segments and plot their corresponding traffic speeds for the time period from 9:30AM to 17:30PM on one experiment day from three sources for comparison in Fig. 5.5. Specifically, we compare our estimated traffic speed of automobiles (s_A), traffic speed (s_T) from the official transport authority, and the Google Maps' indicators on the two road segments. For s_A and s_T, we plot their average speeds with a time window of 15 mins. Figure 5.5 shows that Google Maps only provide rough traffic levels while both s_A and s_T can provide more fine-grained information.

By comparing s_A and s_T in Fig. 5.5, we find some interesting relationships between them. We categorize the traffic speeds of s_A into 3 levels, i.e., low-speed (<45 km/h), medium-speed ($40 \sim 50$ km/h) and high-speed (>50 km/h), for the following comparison between s_A and s_T. When s_A falls in the low speed level, we can see s_A matches s_T well. When s_A is in the high speed level, a gap exists between s_A and s_T. The reason could be that s_A and s_T are derived from bus travel data and taxi travel data, which behave quite differently. Buses are normally slower while taxis travel more aggressively. Though we have transformed bus information to general automobile speed s_A, the taxi speed s_T is still much higher in light traffic scenario. However, both s_A and s_T have the similar variation pattern as shown in Fig. 5.5.

5.4.3   System Overhead

The Goertzel algorithm [1] dominates the computation overhead on mobile phones. In principle, the Goertzel algorithm is computationally efficient than the FFT algorithm for frequency extraction. The complexity of Goertzel algorithm and FFT is $\mathcal{O}(K_g N M)$ and $\mathcal{O}(K_f N \log N)$, respectively, where K_g and K_f are the "cost of operation per unit", M is the number of measured frequencies, and N is the sampling values. In general, the factor K_f is often much larger than K_g [1]. Thus when M is smaller than $\log N$, Goertzel algorithm significantly outperforms FFT. By setting the sampling rate of microphone as 8 kHz for bus detection, the Goertzel algorithm saves more than 60 mW power than FFT for data collection.

We measure the power consumption of two types of mobile phones (i.e., HTC Sensation and Nexus One) under different sensor settings using the Monsoon power monitor. For each setting, we switch off the mobile phone screen, and record the consumed energy over a period of 10 mins and calculate the average power consumption as $\frac{energy}{time}$. Both average power consumption and relative standard deviation (in the parentheses) are reported in Table 5.2. We take the power consumption when no sensors are activated as the baseline. We can see that sampling cellular signals consume negligible power when compared to the baseline, e.g., 72 and 71 mW for HTC phone, respectively. Since mobile phones always maintain connections to nearby cell towers to support telephone calls and SMS service and thus sample the cellular signals at a high frequency, i.e., 1 Hz. As a result, our system takes the almost free lunch and actually introduces marginal energy consumption. The average power consumption for GPS tracking at a sampling rate of 0.05 Hz is as high as 304 mW for HTC and 333 mW for Nexus One. Sampling one GPS signal every 20 s is already very low for the vehicle tracking [8]. The trip data collection via cellular signals only consumes 182 mW for HTC and 196 mW for Nexus One in total. The power consumption, how-

Table 5.2 Power consumption comparison (in mW)

Sensor settings	HTC sensation	Nexus one
No sensors	71 (6)	84 (5)
Cellular 1 Hz	72 (6)	85 (8)
GPS 0.05 Hz	304 (32)	333 (41)
Cellular+Mic(Goertzel)	182 (20)	196 (22)
GPS+Mic(Goertzel)	447 (45)	443 (57)
Cellular+Mic(Goertzel)+Acc	191 (24)	203 (25)

ever, is as high as $\sim$450 mW if we replace cellular signal with GPS signal for the bus trip tracking. Notice that the microphone on the mobile phone has to be kept always on for bus detection no matter the cellular signal or GPS signal is used for vehicle tracking. As our system uses accelerometer on mobile phones to continuously filter out noisy data, we also include the power consumption of accelerometer. The overall power consumption of data collection using our App becomes 191 mW for HTC and 203 mW for Nexus One, with slight increase.

5.5 Summary

This chapter presents the design, implementation and evaluation of a crowdsensing based urban traffic monitoring system, which leverages public buses to probe traffic conditions. The system decomposes traffic sensing tasks to participatory bus riders by utilizing lightweight sensing resources from their mobile phones, and exploits bus route constraints and bus stop references to derive the traffic map. We implement our system and conduct evaluation of 2-month period. The results demonstrate the effectiveness and feasibility of our system for urban traffic monitoring.

References

1. Beck, R., Dempster, A.G., Kale, I.: Finite-precision Goertzel filters used for signal tone detection. IEEE Trans. Circuits Syst. II: Analog Digit. Signal Process. **48**(7), 691–700 (2001)
2. Chakroborty, P., Kikuchi, S.: Using bus travel time data to estimate travel times on urban corridors. Transp. Res. Rec. **18–25**, 2004 (1870)
3. Hemminki, S., Nurmi, P., Tarkoma, S.: Accelerometer-based transportation mode detection on smartphones. In: ACM SenSys (2013)
4. Janecek, A., Valerio, D., Hummel, K.A., Ricciato, F., Hlavacs, H.: The cellular network as a sensor: from mobile phone data to real-time road traffic monitoring. IEEE Trans. Intell. Transp. Syst. **16**(5), 2551–2572 (2015)
5. Lin, K., Kansal, A., Lymberopoulos, D., Zhao, F.: Energy-accuracy trade-off for continuous mobile device location. In: ACM MobiSys (2010)
6. Liu, Z., Jiang, S., Zhou, P., Li, M.: A participatory urban traffic monitoring system: the power of bus riders. IEEE Trans. Intell. Transp. Syst. **18**(10), 2851–2864 (2017)
7. Pu, W., Lin, J., Long, L.: Real-time estimation of urban street segment travel time using buses as speed probes. Transp. Res. Rec. **2129**, 81–89 (2009)
8. Thiagarajan, A., Ravindranath, L., Balakrishnan, H., Madden, S., Girod, L.: Accurate, low-energy trajectory mapping for mobile devices. In: USENIX NSDI (2011)
9. Xie, Y., Xiong, J., Li, M., Jamieson, K.: mD-Track: leveraging multi-dimensionality for passive indoor Wi-Fi tracking. In: ACM MobiCom (2019)
10. Yang, B., Kaul, M., Jensen, C.S.: Using incomplete information for complete weight annotation of road networks. IEEE Trans. Knowl. Data Eng. **26**(5), 1267–1279 (2014)
11. Zhou, P., Zheng, Y., Li, M.: How long to wait?: predicting bus arrival time with mobile phone based participatory sensing. In: ACM MobiSys (2012)
12. Zhu, Y., Li, Z., Zhu, H., Li, M., Zhang, Q.: A compressive sensing approach to urban traffic estimation with probe vehicles. IEEE Trans. Mob. Comput. **12**(11), 2289–2302 (2012)

Chapter 6
Conclusion and Future Work

Abstract In this chapter, we firstly summarize the main content in the book, and then discuss several interesting future research directions.

Keywords Privacy · Data sharing · Multi-source data · Traffic modeling · Parallel computing · Deep learning

6.1 Conclusion

In this book, we introduce the wide available mobility data in the urban cities and its enabled applications for the smart city, i.e., human mobility modeling, urban traffic, green transport, and location based services. In particular, considering the importance of accurate and timely traffic conditions for various smart city applications, we focus on the mobility data driven urban traffic monitoring, and comprehensively review existing works on urban traffic monitoring.

In Chap. 2, we formally describe the problem of mobility data driven urban traffic monitoring, and present a generic framework for mobility data based urban traffic monitoring, which consists of sensing layer, data layer, modeling layer, and application layer. For each layer of the proposed framework, we depict its functionality and widely used techniques or models. Based on this framework, we introduce three different urban traffic monitoring approaches in Chaps. 3–5, respectively. The three approaches differ from each other on both the specific methodology of mobility data collection and the adopted traffic modeling strategy.

In Chap. 3, we propose a compressive sensing based urban traffic monitoring approach. We explore the traffic correlation among road segments with massive mobility data, and propose to model such a relationship via the MLR models. The built MLR models enable us to construct a representation base that can sparsely represent the traffic conditions of all road segments, and provides us an opportunity to recover the complete traffic conditions from sparse mobility data via the compressive sensing theory. We depict the details on building such a traffic condition recovery system, and evaluate its performance with real mobility data.

In Chap. 4, we propose a dynamic correlation modeling based urban traffic monitoring approach. In particular, we argue that more advanced traffic models, e.g., ANN, are better on capturing the complex traffic correlation among road segments. To attack the issues caused by data sparsity and dynamic traffic sensing, we propose a dynamic ANN modeling strategy, and adapt our traffic modeling to the graph-parallel processing framework built on clustered machines. We implement the proposed approach in Apache Spark, and optimize its performance via a set of customized designs. Experiments based on large-scale mobility data demonstrate that our approach can achieve high accuracy in a short time.

In Chap. 5, we propose a crowdsensing based urban traffic monitoring approach. In addition to mobility data provided from official authority or third parties, our system exploits the power of bus riders' mobile phones to participatory sense the urban traffics. We present a holistic design, which includes the lightweight data collection at bus rider's mobile phone and the efficient data analysis on the crowdsourced data for achieving accurate traffic estimations. We implement a prototype of our system and conduct real-world experiments in a testing area. The experiment results demonstrates the feasibility and effectiveness of our system.

6.2 Future Research Directions

With the richer and more diverse mobility data in the urban cities and the emerging of deep learning theory [1], we believe the mobility data driven urban traffic monitoring can be further enhanced by research efforts devoted to the following directions.

(1) Privacy-preserved mobility data sharing and utilization. The majority of mobility data in the urban cities are collected from urban citizens, while mobility data usually contain individual sensitive information, e.g., instant locations or users' mobility patterns. For example, we find that we can learn a driver's driving behavior and habits given a set of her historical GPS trajectories by exploiting the Generative Adversarial Network model [3]. As a result, it raises the concern of privacy issue when urban citizens directly or implicitly share their mobility data with the official authority or the third parties. Therefore, it is necessary and important to devise a mobility data sharing strategy, which can not only preserve users' individual privacy information, but also retain the data utility for effective smart city applications [8]. In particular for the urban traffic monitoring applications, we only need partial information from the mobility data, e.g., location, time and speed, and thus some anonymized or aggregation methods can be used to preserve users' privacy. However, we should devise customized data sharing mechanism that can preserve user's privacy while maximizing the data utility according to the specific application.

(2) Joint optimization of multi-source data and traffic modeling. Various types of mobility data can serve as traffic modeling's input to learn the hidden traffic correlation and characteristics. Traffic conditions may not be reliably inferred from any individual single-source mobility data, while how to select the most appropriate mobility data sources to satisfy the requirements of traffic monitoring is so far

unknown yet [7]. In addition, even such a selection could be eventually figured out, how to further determine suitable traffic model details to fuse these mobility data sources and meanwhile link the input and output are non-trivial neither. Thus, multi-source mobility data driven traffic monitoring encounters a joint optimization of data modality, model structure, and fusion methodology three aspects. One possible solution we propose is to exploit all available mobility data sources for traffic monitoring based on a multi-model strategy. Such a multi-model based traffic estimation is feasible and attractive, where we can exploit the ensemble learning theory [5] to integrate those models and their estimations for a better result.

(3) Parallel computing promoted deep learning to accelerate traffic modeling. In recent years, deep learning [1] has drawn much attention due to its remarkable capability to automatically extract features from large-scale raw data, and has already been successfully applied in various domains. Some pioneer studies also apply deep learning models in the transportation domain [2, 4] and mobility data analysis [6]. To fully extract abstractions from mobility data, deep learning models are usually designed to contain hundreds to thousands of layers and thus numerous parameters need to be tuned. Conventional computing systems are thus inadequate to such computationally intensive tasks. It becomes even more serious when multi-source mobility data are involved, where the storage and computation overheads will significantly increase. Therefore, scalable and efficient parallel computing systems (e.g., computer cluster) arc prcfcrablc to store such big data and accelerate data processing, traffic modeling and the estimations. It is highly expected that a deep learning job can be partitioned into a series of tasks running at different machines in parallel. However, how to achieve the best modeling performances while maintaining the minimum costs on both communications and computations is quite difficult.

References

1. LeCun, Y., Bengio, Y., Hinton, G.: Deep learning. Nature **521**(7553), 436–444 (2015)
2. Liu, Z., Li, Z., Wu, K., Li, M.: Urban traffic prediction from mobility data using deep learning. IEEE Netw. **32**(4), 40–46 (2018)
3. Liu, Z., Zheng, J., Gong, Z., Zhang, H., Wu, K.: Exploiting multi-source data for adversarial driving style representation learning. In: DASFAA (2021)
4. Nguyen, H., Kieu, L.-M., Wen, T., Cai, C.: Deep learning methods in transportation domain: a review. IET Intell. Transp. Syst. **12**(9), 998–1004 (2018)
5. Sagi, O., Rokach, L.: Ensemble learning: a survey. Wiley Interdiscip. Rev.: Data Min. Knowl. Discov. **8**(4), e1249 (2018)
6. Wang, S., Cao, J., Yu, P.: Deep learning for spatio-temporal data mining: a survey. IEEE Trans. Knowl. Data Eng. (2020)
7. Zhang, D., He, T., Zhang, F.: National-scale traffic model calibration in real time with multi-source incomplete data. ACM Trans. Cyber-Phys. Syst. **3**(2), 1–26 (2019)
8. Zheng, X., Cai, Z.: Privacy-preserved data sharing towards multiple parties in industrial IoTs. IEEE J. Sel. Areas Commun. **38**(5), 968–979 (2020)

Index

© The Author(s), under exclusive license to Springer Nature Singapore Pte Ltd. 2021   69
Z. Liu and K. Wu, *Mobility Data-Driven Urban Traffic Monitoring*,
SpringerBriefs in Computer Science,
https://doi.org/10.1007/978-981-16-2241-0

Printed in the United States
by Baker & Taylor Publisher Services